WITHDRAWN

SUNSETS, TWILIGHTS, AND EVENING SKIES

Sunsets, twilights, and evening skies

ADEN *and* MARJORIE MEINEL

Optical Sciences Center
University of Arizona

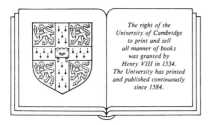

The right of the
University of Cambridge
to print and sell
all manner of books
was granted by
Henry VIII in 1534.
The University has printed
and published continuously
since 1584.

CAMBRIDGE UNIVERSITY PRESS

CAMBRIDGE
NEW YORK   PORT CHESTER
MELBOURNE   SYDNEY

Published by the Press Syndicate of the University of Cambridge
The Pitt Building, Trumpington Street, Cambridge CB2 1RP
40 West 20th Street, New York, NY 10011, USA
296 Beaconsfield Parade, Middle Park, Melbourne 3206, Australia
10 Stamford Road, Oakleigh, Melbourne 3166, Australia

© Cambridge University Press 1983

First published 1983
First paperback edition 1991

Printed in the United States of America

Library of Congress Cataloging in Publication Data
Meinel, Aden B.
Sunsets, twilights, and evening skies.
Includes bibliographical references and index.
1. Meteorological optics. I. Meinel, Marjorie Pettit.
II. Title.
QC975.2.M44 1983   551.5′66   83-1794

ISBN 0-521-25220-2 hardback
ISBN 0-521-40647-1 paperback

Plates 2-3, 3-1, 3-5. D. J. K. O'Connell and C. Treusch, Specola Vaticana, Vatican City State.
See Richershe Astronomiche 4 (Specola Vaticana, 1958), also *The green flash and other low sun phenomena* (North Holland, Amsterdam, 1958).
Plates 3-2c, 3-3, 3-4. Dennis diCicco, *Sky and Telescope Magazine*.
Plate 4-13. Agnes Meller, Caracas, Venezuela. Color prints taken in Caracas, Venezuela, at about 5:30 P.M. looking due east.
Plate 7-3. NASA Langley Research Center (M. P. McCormick).
Plate 9-1. Photo by Yngvar Gotaas, Geophysical Institute, University of Alaska, Fairbanks, Alaska.
Plates 13-1 to 13-3, 13-5 to 13-9. NASA Johnson Space Center (Owen Garriott).
Plate 13-4. M. Ackerman, C. Lippins, and M. Lechavallier, Institut d'Aeronomie Spatiale de Belgique, 3 Avenue Circulair, B1180 Brussels, Belgium. See *Nature 287,* 614 (1980).
Plates 13-10, 13-12. Provided through the courtesy of NASA Jet Propulsion Laboratory, California Institute of Techology, Pasadena, California.

השמים מספרים כבוד-א-ל ומעשה ידיו מגיד הרקיע

The heavens declare the glory of God and the vault of heaven shows
the work of his hands.   PSALM 19:1

Contents

Preface		*page* ix
1	Panorama	1
2	Sunset prelude	5
3	The green flash	19
4	The earth's shadow and sunset phenomena	29
5	Volcanic eruptions	39
6	Volcanic twilights	51
7	Twilight science	63
8	Bishop's ring and blue suns	79
9	Noctilucent clouds	87
10	Zodiacal light and the gegenschein	91
11	Light of the night sky and the aurora	101
12	City lights	115
13	Twilight on the planets	121
14	Celestial visitors	133
15	Reflections	149
Notes		157
Name Index		159
Subject Index		161

Preface

A word may be useful to the reader regarding our attempt to unite in this book a scientific overview with the aesthetics of a subject that is of intrinsic beauty. The thread that ties this book together is our personal experiences over many years. You will notice that we relate our growing awareness of new aspects of the sky through personal stories and stories our friends have told us. This style of writing is not the norm for science today, which seems to delight in using chilly prose strewn with words known only to a small inner circle. Perhaps we are old-fashioned in this regard: We love to read the scientific journals from the turn of the century. You will find herein generous quotes from some concerning the 1883 eruption of Krakatoa and the subsequent optical phenomena. Science then was written with a full sense of excitement. Many articles were written or contributed by nonscientists and especially by missionaries, people hardly qualified to enter the lists of today's modern science journal; yet their articles were filled with accurate information well expressed. We hope the spirit of this book will be in the old fashion, when science was open to every reader as a pleasant way to learn the fascinating facts of nature while seated quietly at home.

We have followed the skies all our lives wherever we have happened to be. Our collection of color pictures and notes had accumulated for many years. A pause resulting from health events gave us the time to reflect on past experiences and form the resolve to share them with others through a book. Writing a book is more than simply collecting what has been casually gathered over a span of years. One quickly finds gaps in the topics that need to be filled. We resolved to fill some gaps – for instance, in regard to the zodiacal light and the counterglow. We had the time and we also at last had an all-sky camera lens of sufficient speed to record the evening sky with the sensitivity needed to visualize such faint but striking phenomena.

Preface

This book is written for the many inquisitive people who enjoy nature and are interested in knowing the spectrum of appearances of sunsets and evening skies. Younger scientists who are using such modern tools as lidar and high-altitude aircraft to explore the nature and consequences of atmospheric disturbances also can benefit from our less-precise observations. When one knows the appearances of the skies, one can easily detect the difference between a normal sunset and one enhanced in subtle but beautiful ways by either nature or human activities. To illuminate facets of interest to the amateur scientist or student we occasionally show some mathematical relationships in graphical form.

This book has profited from the interest of many friends, some of whom have contributed their favorite pictures. Many more illustrations were proffered than could be fitted within the budget set for publication costs.

We hope you will enjoy our book and look for the things we describe when you, too, view the sunsets, twilights, and evening skies.

Aden and Marjorie Meinel

Tucson, Arizona
1983

I

Panorama

Passages. The word conjures up images of travel – of going from one place to another. The Age of Exploration was a time of many passages, of ships sailing around the Horn of Africa into the Indian Ocean or threading their way through the Strait of Magellan out into the vast Pacific. This book deals with quite a different type of passage, a fixed passage wherein it is the majestic turning of the earth that brings ever-changing scenes to you wherever you may be.

Passages stir the emotions. Our days are filled with passages: in our personal lives, in the seasons, and – one of the most beautiful – in the diurnal passage from day into night followed by night into day. In this book we share with you our experiences with many different types of sunset, twilight, and night phenomena along with their scientific explanations in order better to attune your appreciation of what you, too, may encounter.

We have spent most of our lives in and near the desert, but we also have traveled to many countries in many parts of the world. Wherever we are, we note the appearance of the sky and the sunsets and sunrises, each made individual by foreground settings as well as sky conditions. Even the expanse of a cloud-free sunset sky bears the signature of the location. Scientists have for years studied these clear sunsets and twilights to learn more about the atmosphere – the lower atmosphere as well as its upper limits. We will not dwell on these scientific studies, but will cite in the Notes at the end of the book references where an interested reader may pursue to the fringes of knowledge.

There is a special feeling of serenity when we stand on the hillcrest above our Tucson home and look westward beyond the silhouette of the angular Tucson Mountains to where, as in Bing Crosby's theme song of Big Radio days, "the blue of the night meets the gold of the day." The soft desert breeze whispering through the spines of the lofty saguaro cactus sets an incomparable scene, holding our thoughts even as the stars become visible through the fast-fading twilight. Astronomical twilight has ended and night has begun.

In recent years, since the autumn of 1963 to be exact, the world has been treated to especially colorful transitions between day and night. In March of 1963 a tremendous volcanic eruption of Mount Agung on the South Pacific isle of Bali spewed forth an immense volume of ash and gas. The atmosphere spread this veil over the entire earth, reaching northern latitudes in late September. We noticed that after the fading of the usual sunset a second sunset appeared, a fiery red glowing stratum against which lower clouds were darkly silhouetted. We remembered years ago reading in geology class that the world had experienced "vivid sunsets" after the massive 1883 eruption of Krakatoa on Java. Were we having a repeat show in 1963? What, in fact, were the descriptions of the 1883 sunsets? Our curiosity was pricked and we headed for the library, knowing only the name of the volcano and the year of its eruption. What we found is related in Chapters 5 and 6 of this book.

The 1963 eruption of Agung signaled the onset of "volcanic twilights," but the brilliance slowly faded, only to be reborn, enhanced by subsequent eruptions of other volcanoes. What must the sunset skies have been in geologic ages past when periods of intense mountain building must have been accompanied by far more frequent eruptions of greater violence? Did the veil of dust and newly emitted gases even change the climate? The geologic record of temperatures visible today in the permanent record of oxygen isotope abundances in limestone and chert layers shows that the atmosphere did go through vast changes, both warmer and colder. In fact, the earth's temperature today is almost as low as in the ice ages, but in some past epochs it has been very hot. Volcanoes put out veils of ash that can lower the earth's temperature. At the same time they put out gases. The ash veil soon falls to earth, but the gases persist for many years. The resultant "greenhouse" effect can and will heat the earth, driving with it the course of life on this beautiful planet. Knowledge of this awesome reality gives these recent volcanic twilights a fascination for us. We are seeing the faint reflection of ages past, before the advent of mankind, where there was no one to see and appreciate the beauty of the passages from day into night.

What are the phenomena to be observed as day passes into night? First there is the prelude to sunset. As the sun lowers toward the horizon, it changes color and becomes less brilliant. In urban surroundings it may redden and even disappear in the gloom of haze, yet the colorations of sunset still can tint the gloom. Even in the sharp desert air the sun noticeably changes color and even dims, although a motorist driving westward at the close of day may still be blinded by the setting sun and wish for the soothing smog absorption of a Los Angeles afternoon.

The setting sun becomes flattened by atmospheric refraction. In a wet, stable atmosphere, as often occurs along the Pacific coast, the sun takes on the bulbous, stratified appearance of a Chinese lantern, constantly changing shape as it sinks toward the horizon through differing layers of the atmosphere.

Have you noticed the long blue shadow bands that sometimes spread accross the sky, converging to the setting sun? Or, at the same time, the occasional blue rays converging in the pale roseate east? Finally the sun itself slips below the horizon, and at the last moment the rare green flash may appear. This flash occurs under several conditions frequently met in such climates as the desert or the Hawaiian Islands.

The normal clear-sky sunset has its own sequence of coloration. As in the song, the blue of the night does meet the gold of the day as the golden arch sets in the west and the eastern blue has swept westward across the sky and covered all but a yellowish horizon band. The stars appear one by one as twilight comes to an end. If the night is very dark, a faint wedge of luminosity resembling the Milky Way can be seen in the west with its apex pointing along the ecliptic (the annual pathway of the sun). This is the zodiacal light, sunlight reflected from interplanetary dust – what remains of the cosmic material from which the planets were formed.

Sometimes and at some places illuminated clouds can be seen even though the sun has been set for more than an hour and the sky is very dark. These noctilucent, or glowing, clouds far above the stratosphere are rarely seen except in high latitudes at certain times of year. If you live downwind from a missile launch site you probably have seen a manmade noctilucent cloud, sunlit and also at a very high altitude above the earth.

The glowing aurora, although caused by plasma from the sun by way of the Van Allen belt of ions about the earth, displays its sunlit effects at sunset and sunrise.

Even after the zodiacal light has disappeared, subdued effects of the full moon sometimes are visible. Even on the blackest of nights in the middle of the night a few observers have been able to see at the antisolar point a faint hazy patch of reflected sunlight from the interplanetary dust outside the earth's orbit. This is the gegenschein, and in Chapter 10 we describe how we tried to photograph this elusive counterglow.

We hope that after perusing this book you will be able to see many of these phenomena and with enhanced understanding enjoy their beauty.

2

Sunset prelude

Before turning to the beautiful phenomena associated with sunrises and sunsets, we will look at some geometrical consequences of the daily and annual path of the sun combined with the latitude of the observer. We will also define the terms applied to this geometry because they will occasionally appear in later chapters.

PATH OF THE SUN

The sun follows a well-defined path across the sky each day, a path that sweeps from east to west with the eastward rotation of the earth. Each day, however, the sun is in a slightly different place with reference to the stars, which daily outspeed the sun by 4 minutes and 55 seconds. Thus the sun completes its annual journey in about 365.25 days to return to the same place against the background stars. This slow march of the sun through the celestial sphere is along a path called the ecliptic, a sinusoidal path reaching its northernmost point on 21 June and its southernmost on 21 December (the solstices).

The daily sweep of the sun crosses the meridian (the imaginary line from the south point of your horizon through the zenith and the north point) at an angle above the horizon that depends on your latitude. On the equinoxes this angle is 90° minus your latitude. It is largest at the equator, being 90°, which means that the sun passes through the zenith. Your shadow falls at your feet at noon on these two days at the equator. At the 41° latitude of New York City this angle is 49°, and you cast a noon shadow about equal to your height.

Latitude makes a big difference in the duration of sunset effects. At the rising and setting points of the sun the diurnal path is tilted with respect to the vertical by the angle of the latitude of the observer. At the equator this angle is zero, and the sun plunges quickly below the

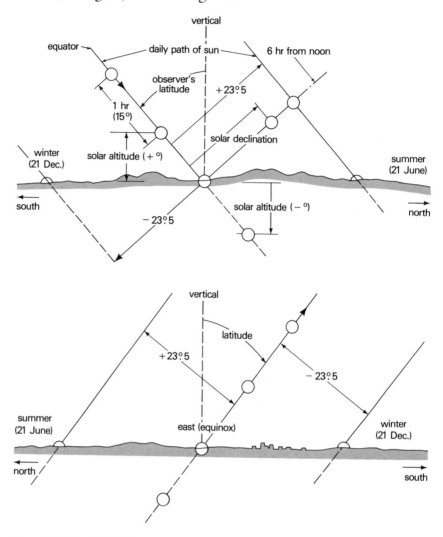

Figure 2-1. Top: Path of the setting sun at the equinoxes and solstices. Bottom: Path of the rising sun. Compare this diagram with the time-lapse photograph in Plate 2-12.

horizon because the solar altitude is decreasing at its greatest rate. At high latitudes this angle becomes large, and the sun sets more slowly because the solar altitude is decreasing slowly. This makes twilight pass into night rapidly in the tropics and slowly in high latitudes, even though the sun moves along its diurnal path at essentially the same rate. This geometry is shown in Figure 2-1.

It must be remembered that the daily path of the sun is not represented by the plane geometry depicted in Figure 2-1; such a representation requires spherical geometry. This distinction is not important for low latitudes during the period of twilight, but is very important

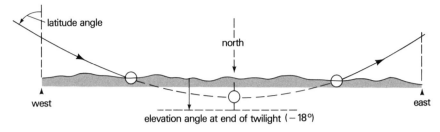

Figure 2-2. Path of the sun in summer at a northern latitude, where it is not quite dark at midnight.

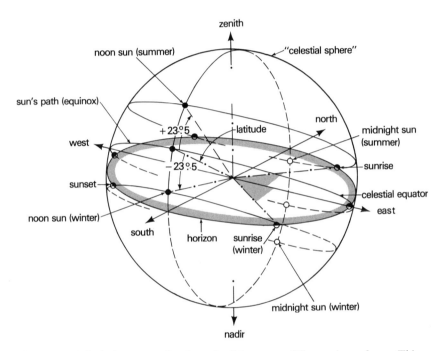

Figure 2-3. Spherical geometry for the path of the sun at different times of year. This example is for a rather far northern latitude, where the sun barely rises in winter and hardly sets in summer.

at high latitudes as one approaches the Arctic or Antarctic. In summer the daily path of the sun dips only briefly, or not at all, below the horizon. If you look toward the north point, the actual path in summer would appear as in Figure 2-2. The sun may reach its lower culmination at only a small solar altitude angle below the horizon so that even at midnight the sun is still shining on the upper atmosphere illuminating, for example, the polar noctilucent clouds. Figure 2-3 shows the correct spherical geometry for the path of the sun at different times of year for a midlatitude observer.

Table 2-1. *Time of sunset and sunrise, in local sun time, for various latitudes at the equinoxes and solstices*

	Winter solstice			Equinox			Summer solstice		
	30°	40°	50°	60°	30°	40°	50°	60°	30°	40°	50°	60°
Sunset (P.M.)	5:02	4:34	3:55	2:44	6:00	6:00	6:00	6:00	6:58	7:26	8:05	9:15
	17:02	16:34	15:55	14:44	18:00	18:00	18:00	18:00	18:58	19:26	20:05	21:15
Sunrise	06:58	07:26	08:05	09:15	06:00	06:00	06:00	06:00	05:02	04:34	03:55	02:44

The consequence of this geometry is that the time of the end of twilight and the beginning of astronomical night is quite different for different latitudes. The end of twilight is generally considered to be when the sun reaches an angle of about 18° below the horizon.

TIME OF SUNSET AND SUNRISE

The time of sunset or sunrise also varies throughout the year because of the geometric facts described in the preceding section. Table 2-1 gives these times for several latitudes at the equinoxes and solstices. It should be noted that these times are "sun time," not local standard time. Because the sun passes through 360° of longitude each 24 hours, a 1-hour time zone covers 15° of longitude. There are places where the local civil time is far off sun time. For instance, China extends over 45° of longitude, but has only one time zone. In Shanghai civil time coincides with true sun time; in the western Takli Makan desert it is 3 hours later than true sun time. In addition, some parts of the world add an hour in the summer as daylight-saving time, which makes the actual clock times for sunset and sunrise 1 hour later.

A graphical description of how the time of sunset varies with summer and winter solstices is shown for different latitudes in Figure 2-4.

ABSORPTION AND REDDENING

Light from the setting sun must pass through an increasing mass of air, which causes the sun to dim and become reddened. The amount of coloration depends on the air itself or, rather, on the aerosols and particulate matter suspended in the air. This change in brightness and color is shown in the time-sequence photograph in Plate 2-1. The sun images were taken 5 minutes apart through a dense filter with a clear sky until near sunset. The filter was removed and a normal exposure added after sunset to show the cirrus clouds.

The absorption and reddening are caused by the combined effects of true absorption by water vapor and ozone and of scattering by atmospheric gases, aerosols, and dust. Aerosols are small aggregates of gases and liquids condensed about tiny condensation nuclei such as dust particles. They are so small that they remain suspended in the atmosphere indefinitely. They tend to grow slowly and precipitate out along with grosser dust particles, but are continually being replaced by newly formed aerosol particles.

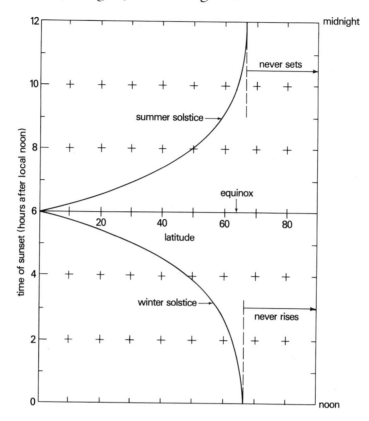

Figure 2-4. Graphical description of how the time of sunset varies at summer and winter solstices for different latitudes. For time of sunrise, simply reverse the numbers on the vertical axis.

Water vapor causes true absorption in certain wavelength bands of the spectrum, sometimes called the rain bands. Since the rain bands absorb mainly in the red, heavy water concentration tends to render the sky greenish. Ozone has weak diffuse bands in the green and stronger ones in the ultraviolet. The main reason why the setting sun is reddened is that most aerosols and atmospheric gases scatter blue light more strongly than red light. The red light from the sun penetrates to your eye; its blue light is lost in producing the blue sky coloration. Ozone also contributes to this effect.

Scattering by a pure gas such as nitrogen, which constitutes about 80 percent of our atmosphere, is strongly wavelength-dependent, increasing steadily toward the blue. It is called Rayleigh scattering after Lord Rayleigh, who first gave a mathematical explanation of the effect. The normal blue sky is caused about half by Rayleigh scattering and half by aerosol and ozone scattering and absorption. In a desert

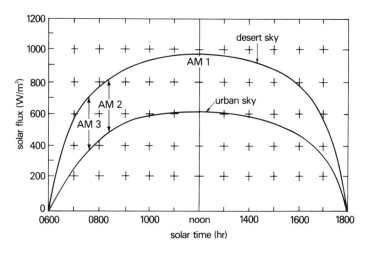

Figure 2-5. Graphical description of how the solar brightness (intensity) varies with time of day for a typical desert sky and for a typical urban sky. Air masses 1, 2, and 3 are noted as AM1, AM2, AM3.

climate little water vapor is aloft and a minimum of aerosols, so the sky tends to be a pure blue. In urban climates the sky tends to whiten, with the blue tint subdued in comparison with a desert sky.

During the day horizontal visibility is a good indicator of the amount of reddening to expect at sunset that evening. From our home we can usually see mountains, such as Baboquivari, which are 80 km (50 mi) distant, and from the Kitt Peak National Observatory we can see mountain ranges 300 km (250 mi) away. On such an evening the sun appears hardly reddened at sunset. On the other hand, on a hazy day in Los Angeles one can sometimes see scarcely 2 km (1.4 mi), and a heavily reddened setting sun is assured, the pearly gray sky even obscuring the sun well before it reaches the horizon.

A graph of the variation of the solar intensity from sunrise to sunset shows a strongly convex curve of similar shape for both a clear sky and a somewhat hazy sky (Figure 2-5). This roll-off in intensity also clearly shows in Plate 2.1, but to the eye the sun seems equally brilliant all day until close to sunset. Anyone who has driven west at the equinox in the desert can testify to the blinding brilliance of the sun right up to sunset. The reason for this difference between a photograph and the eye is that the eye responds logarithmically and film, linearly. A visual change of one step in brilliance is more like a change of ten steps on film. This is one reason why an exposure meter is so helpful for getting a good photograph.

This automatic adjustment of the eye to the brilliance of the scene can cause another instance of misjudgment: the expected performance of a solar collection device. A hazy or overcast day may seem almost as bright as a clear one. This illusion is in part because the sky you see before you while going about your daily business is that part near the horizon. The clear-sky sun is bright and shadows are crisp, yet the very blue horizon sky itself is not very bright. On an overcast day the shadows are almost missing, but the whitish horizon sky is bright. On the clear day your solar device works fine, but on the overcast day you may not collect much usable energy because, like photographic film, the solar collection device is relatively linear compared with your eye, which is logarithmic. We will have further opportunity to encounter the remarkable logarithmic response of the eye in Chapters 4, 9, and 10.

REFRACTION

Many optical effects are a result of the passage of sunlight through the air. The refractivity of air (1.0003) is slight compared with that of glass (1.5000); even so, it is sufficient to cause the horizon sun to appear to be its whole diameter above its true position, as illustrated in Figure 2-6. Terrestrial dimensions on the horizon are so large that this small bending of only $0°.5$ can cause some interesting effects, such as the green flash which will be discussed in Chapter 3. Figure 2-7 shows the change in the angle of refraction with apparent altitude angle of the object. The amount of refraction differs with color. Through a blue filter the sun appears slightly above its position as seen through a red filter.

The curvature of the earth makes the amount of refraction depend also on where you are situated. If you are flying at an altitude of 10 km (33,000 ft) and see the sun or moon on the horizon, refraction effects are significantly increased over what you see when standing at sea level. An astronaut orbiting far above the atmosphere would see twice the effect because his line of sight passes twice through the atmosphere. This augmented refraction will be discussed in more detail in Chapter 13.

The earth's atmosphere is dense at sea level, rarifying rapidly with altitude. Refraction depends on density, so it rapidly decreases with altitude. In a strict mathematical sense this diminution of refraction with altitude and with temperature of the air layers does not matter as long as the layers are horizontal and flat. All that matters is the density

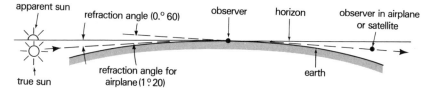

Figure 2-6. Geometry of atmospheric refraction for an observer on the earth's surface and one in an airplane or satellite. Note that the double passage of the solar ray through the atmosphere doubles the refraction angle for the observer aloft.

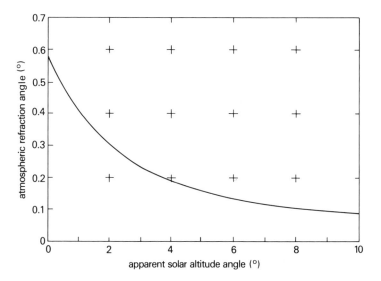

Figure 2-7. Graphical description of how the refraction angle varies with the apparent altitude angle of the sun above the horizon.

and temperature at your location. The density and temperature in the layers above you do not affect the angle of refraction for you.

We saw an ingenious use of this mathematical fact during our visit to the Nanjing Astronomical Instruments Factory in China, where our friend, Vice Chief Engineer Hu Ningsheng, builds instruments to measure star positions precisely. To eliminate the effect of refraction on the true position of stars, he placed the optics of the instrument inside an enclosure and removed the air. The instrument looks out at the stars through a horizontal glass window. Because there is no air inside, and hence no refractivity, theoretically there will be no effect of the atmosphere above the glass window, and the stars will appear in exactly their correct positions. And, in fact, they do.

OBLATENESS

Refraction causes an interesting change in appearance of the setting sun. The casual observer sees that the sun is perfectly circular when high in the sky, but the camera shows the sun is flattened as it nears the horizon; so also is the full moon. If you look at the full moon or the sun (through a filter) when it is just above the horizon, it does not appear conspicuously flattened. This is an optical illusion. Another optical illusion enhances the size of the moon near the horizon. This illusion results because your mind subconsciously relates the distant sun or moon to the relatively nearby horizon. You translate the sun's or moon's size vis à vis this apparent terrestrial distance. To eliminate relating the celestial object to the terrestrial scenery, look at it from a different point of view. Plate 2-2 presents two pictures of the solar image taken near the horizon: as you normally see it (left) and turned sideways (right). If you cover the left-hand picture, the right-hand one looks quite elongated. Another way to change your viewpoint is to look at the sunset sky with your head turned upside down. You will increase your awareness of sky colors by this method if you do not mind what onlookers may think. The blue at the zenith will seem bluer and the horizon sunset reds, redder.

A graph is useful to show the amount of flattening seen by a surface observer and by one at high altitude. The oblateness, expressed as percent flattening versus apparent solar altitude angle, is shown in Figure 2-8. The apparent shape of the sun is shown at the right side of this figure. Look for this excess flattening of the sun at sunset next time you are flying at jet altitude. The dimming of the sun through a thick, hazy lower atmosphere will enable you to see the effect easily.

CHINESE-LANTERN EFFECT

Watching the sun set to an ocean horizon, you are aware that its shape is becoming quite different from the smooth ellipse depicted in Plate 2-2. Such a scene is shown in Plate 2-3, which consists of several pictures taken as the sun sets through a stratum of air that has different refractivities. These pictures, taken through a telescope, show laterally symmetrical distortions that make the sun resemble the traditional Chinese paper lantern. Note that as the sun sinks to the horizon the shape changes, but the apparent layering remains fixed above the ocean surface. Note also that although the sun is greatly distorted, the silhouette of the passing ship is sharply imaged. This anomalous refrac-

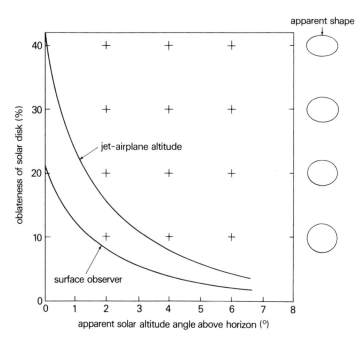

apparent shape

Figure 2-8. Graphical description of the percentage flattening of the sun as a function of altitude angle above the horizon.

tion will enter the discussion of the green-flash phenomenon to be examined in Chapter 3, so we should look closer at the origin of the effect.

We stated in a preceding section that atmospheric refraction depends only on the air temperature and density at the location of the observer, not on the state of stratification in the atmosphere. If that is so, how does the Chinese-lantern effect arise? Quite clearly the distorted image of the setting sun shows the effect of layers of different refractivity. The law relating to the independence of refraction on layering is experimentally accurate for objects high above the horizon. Astronomical observations are generally made with the object as high in the sky as possible to avoid problems arising from the real nature of the atmosphere. For such observations the atmospheric layers are accurately plane-parallel, and the law holds. For the setting sun we must also consider the curvature of the earth.

Looking at the distant horizon in Plate 2-3, we can clearly see stratification. The layer closest to the sea has more absorption, reddening the sun more than the layer directly above. Moreover, the upper boundary of the reddened layer is sharp. This is because we see it edge-on owing to the curvature of the earth. The observer's line of sight to the top of the sun does not penetrate this layer, so the rays are not refracted by it. The line of sight to the bottom, however, does penetrate this layer and is refracted by it by a different amount than is the top. The scientific explanation of this anomalous refraction is that the angle of the ray exiting the layer is different from the angle of the ray entering the layer owing to the curvature of the earth. The invariance law holds only when the angle of the ray exiting the layer is exactly the same as the angle of the ray entering: for example, a plane-parallel atmosphere.

We have never seen the Chinese-lantern effect from our home in Tucson. The desert atmosphere is not strongly layered and is usually well stirred by the desert breezes in the afternoon. We have seen strong stratification down in the valleys at dawn when night-chilled air settles there to form spectacular mirages, but we have never seen the sun through these strong inversions. The Chinese-lantern effect is usually seen along seacoastal regions, such as Los Angeles, where layers of different humidity can form – especially in autumn when stagnant air traps the moisture and smog, making some days quite miserable.

THE NOVAYA ZEMLYA SOLAR MIRAGE

The layering of air of different refractivity also produces a wide range of phenomena of the mirage type. Rays can become trapped within a layer, especially in Arctic regions. One can see the sun when it actually is much farther below the horizon than can be explained by normal refraction. This effect is termed the Novaya Zemlya mirage. The ship of the Arctic explorer Willem Barents was entrapped in polar ice on Novaya Zemlya Land, north of Siberia, during a search for a Northeast Passage to the riches of the Orient. The end of the long winter night was eagerly awaited. At noontime near the end of January 1597, the explorer was astonished to see the rim of the sun appear above the icy wastes to the south. His calculations had led him not to expect the sun to rise until 2 weeks later. On that January noon the sun was about 5° below the horizon, and normal refraction would raise it by only 0°.5.

Barents's report of this welcome sight was doubted by scientists of that day. The astronomer Johannes Kepler did try to find a rational explanation, but was unable to do so. The reality of the Barents observation was confirmed by another explorer, Sir Ernest Shackleton, on his last expedition to the Antarctic continent in 1915. In this case the

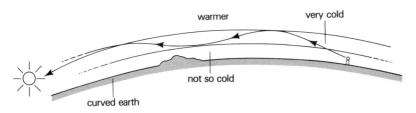

Figure 2-9. Possible origin of the Novaya Zemlya effect in ducting the solar rays trapped in a cold stratum.

upper edge of the sun briefly appeared 7 days after it had set for the winter. Shackleton reported that twice it reappeared.

The Novaya Zemlya effect was studied by two optical scientists, W. H. Lehn and B. German, who went to Canada in May 1979. Because the period of the midnight sun was about to begin, they looked for the reverse effect Shackleton had seen: the sun refusing to drop below the horizon. They found it, describing the remaining limb of the sun as a rectangle of blazing light. The rectangle marked the thickness of the warm-air inversion over the cold terrain. Lehn and I. I. Schroeder suggest that this effect explains the island said in the ancient sagas to lie northwest of Iceland, the island sought by Eric the Red when he was exiled from Iceland.[1] Eric found not some ghostly islands but the ice-shrouded, optimistically named minicontinent of Greenland.

The origin of anomalous refraction can be understood if we consider the interface between a cool and a warm layer to be smooth. In Figure 2-9 a very cold layer is bounded on both top and bottom by warmer air. This cold layer is not of infinite length, but long enough to cause the line-of-sight ray to be "reflected" by refraction several times. In this manner the ray would exit the trapping layer so distant from the observing explorer that he could glimpse sunlight while the sun was still far below his horizon.

3

The green flash

At the last moment as the sun disappears below a distant clear horizon there can be a glint of vivid green light. We say "can" because the phenomenon is elusive, but one worth looking for. What is its history? What are the conditions under which one can expect to see it?

HISTORICAL SIGHTINGS OF THE GREEN FLASH

The historical record is sparse, a fact that led D. J. K. O'Connell, S.J., to remark in his book, *The green flash:* "The green flash appears at times so vivid, even to the naked eye, that it is surprising that the ealiest references to it are of comparatively recent date. One would have expected such diligent observers as the early astronomers of Babylonia, Chaldea, and Egypt would have noted the phenomenon."[1]

It first became a phenomenon of interest during the outburst of observational sensitivity associated with the Victorian Age. The frequency of letters to the science journals from missionaries scattered around the world in the days of colonial empire building shows how these missionaries-doctors-naturalists had become attuned to look carefully at nature, spurred by a new awareness of the diversity on the planet earth. Two notable natural catastrophes undoubtedly drew new attention to the varied appearance of the sun and sky: the eruptions first of Tambora in 1815 and then of Krakatoa in 1883. Tambora was the largest volcanic eruption since the protohistoric explosion of Santorin (c. 1500 B.C.) in the eastern Mediterranean, which wiped out the Minoan civilization at the same time that Moses was being led by the "pillar of cloud by day and fire by night." Krakatoa was a lesser eruption, but it occurred at a time of great scientific sensitivity and hence was more carefully observed. It is the subject of Chapter 5.

The earliest undisputed scientific record of an observation of the green flash was made by W. Swan in 1865 when he saw a "dazzling emerald green" flash at sunrise over a distant mountain. A number of later observers saw it better at sunrise than at sunset, but knowing exactly where to look presented a problem. It should be noted that Swan did not publish his observation until 1883, after the worldwide attention to green suns and other phenomena arising from the Krakatoa event.

Although Swan's observation is generally accepted as the first scientific mention of the green flash, we happened upon one made 31 years previous to his in the narrative of the expedition of H.M.S. *Terror* to the Arctic in 1863–7. The commander of the expedition, Captain Back, reported

In the morning however, at a quarter before ten o'clock [17 January, when the sun barely rose above the south horizon] while standing on an ice hummock about seventeen feet high, and looking toward the east, I had observed the upper limb of the sun, as it filled a triangular cleft on the ridge of the headland, of the most brilliant emerald colour, a phenomenon which I had not witnessed before in these regions.[2]

It has been suggested that Jules Verne's novel, *Le Rayon vert* (*The green ray*), published in 1882, first attracted general attention to the phenomenon. The great British scientist Lord Kelvin (William Thomson) referred explicitly to Verne's novel when he wrote in 1899 of his observation of a "blue flash" just as the sun rose over Mont Blanc, Switzerland. In his memoirs he remarked having earlier, in 1893, seen the green flash at sunset. By 1926, the number of sightings had become so large that the subject even formed a Ph.D. dissertation by P. F. Keuper. Many of the observations were from shipboard sightings, giving a strong clue to the cause of the phenomenon.

The green flash is not often seen, even with a clear view of the sun at a distant horizon; in fact, seeing it is the exception, not the rule. If it were the normal consequence of the atmosphere, one would expect to see it every sunset when the sky looks reasonably clear. The elusiveness of the green ray, however, has disappointed many an eagerly anticipated observation.

In his book, Father O'Connell remarks that during his visit to Lick Obervatory in California he and the astronomer Robert Aitken stood on the west front steps watching the sun setting below the horizon of the distant Pacific Ocean, but seeing no green flash, even though Aitken said he had often seen it. During our student days at Berkeley we too stood at the same location looking in vain for the green flash. Not until we lived in the desert did we see it with any frequency. Aitken wrote that although at some periods he regularly saw the green flash from Lick, he also failed to see it for months at a time.

The capriciousness of the flash led some scientists to suggest it was a purely physiological phenomenon rather than a physical one. Experiments were even performed to demonstrate that it could be due to fatigue of the retina, which caused the complementary color to be seen after the eye had been dazzled by the brilliant orange or red edge of the setting sun. It is true that if you close your eyes after looking at a bright red light the afterimage will be greenish. Those who held to this explanation clearly had never personally beheld the green flash. The afterimage from retinal fatigue is a dull complementary image that only slowly fades after some tens of seconds. The green flash is a brilliant emerald glint that lasts for only a split second before it too follows the set sun below the horizon. The physiological theory also fails to explain the sunrise green flash, which appears before the sun has risen.

In 1887 a curiously interesting theory evolved from the many sightings of the green flash from shipboard: that it was caused by sunlight passing through the crests of waves on the horizon. The green color was said to be due to the passage of light through the green sea water. However, the observation of the green flash over land as well as sea means that this theory is not sufficient or even plausible.

EXPLANATIONS

These explanations have been so often resurrected that O'Connell sympathized with an earlier writer, Sir Arthur Schuster, who wrote in 1922: "The same arguments are repeated over and over again, until we feel that a horse dead and duly flogged had better be buried; this might save us from being worried by its ghosts and reincarnations."[3] What then is the correct scientific explanation of the green flash that accounts for both its manifestations and its conspicuous absences at the same horizon and when the sun is clearly visible at sunset?

Atmospheric refraction is the first place to look for an explanation. In Chapter 2 we saw that refraction causes the sun to appear to be displaced upward by slightly more than its diameter. It is also known that refraction is larger for green and blue light than for red light. This means that the image of the sun viewed in green light will appear to be slightly higher above the horizon than the image in red (Figure 3-1). This effect can readily be seen if one looks through a small telescope

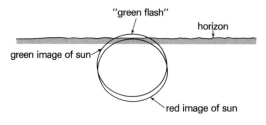

Figure 3-1. Theoretical explanation of the green flash. Atmospheric refraction causes the sun's image to be separated by color, the green image being seen above the red image.

at Venus or a bright star near the horizon, as shown in Plate 3-5. This image of Venus shows different diameters in the red and the blue. This is not an atmospheric effect. The use of a refracting (lens) type of telescope and focus and magnification change with color produce such an effect when the image is greatly enlarged. The main thing to be noted in this photograph is the vertical displacement of the blue image above the red.

The angular diameter of the sun is so large that this effect is less easily visible for the sun than for Venus. A theoretical calculation of the effect can be made based on the measured refractive index of air. The dispersive power from blue (n = 1.000295) to red (1.000292) is only 3 parts in 294, or 1.02 percent. This means that the refraction difference is only $0°.006$, compared with $0°.530$ for the refraction itself. This small angle is only 20 arc seconds; the eye can resolve only about 120 arc seconds.

The situation at the moment of sunset for "red light" is shown in Figure 3-1. The tiny strip of the sun remaining above the horizon is too thin to be seen except as a glint of light that goes through a quick color change. The sun moves $1°.0$ in 4 minutes, so the strip of $0°.006$ passes in 1.4 seconds. If one had ideal conditions – a sharp horizon, perfectly clear air, homogeneous atmosphere, and an expanded time sensitivity to slow the solar motion – one would see the dazzling strip of sun change from normal by the successive subtraction of first red, then yellow, then green, with the remaining glint pure violet-blue. The combined effect of the remaining light during this change is to see progressively a yellowish, then a greenish image, and finally the pure color of the remaining blue. Knowing what to expect, we have clearly seen the progression.

If the "normal" atmosphere can produce a green flash, why then is the flash not always seen under good sunset conditions? In fact, why isn't it a blue flash?

Location can be important if one wished to see a green or blue flash over land. The terrain must be bare of vegetation so that a sharp boundary divides the image of the rim of the sun. Dick Kinnaird wrote us about how he and his wife, Peggy, had often been able to see a green and even a blue flash:

We see the flash as the sun rises over Cadillac mountain on Mt. Desert Island in Maine at altitudes along the profile of about 1° to 2°. On one such occasion the green flash appeared weakly surmounted by a small blue crest, but on three occasions no green appeared after the blue . . . A barren smooth rock face is required. For perhaps a minute or so before the flash the mountain profile in the vicinity [of where the sun will rise] is fringed in a silvery or yellowish glow along the profile.

This fringe is undoubtedly diffraction. With such a stark mountain profile the Kinnairds were able to see the effect of normal atmospheric dispersion as illustrated in Figure 3-1. Being morning, the rock face was also undoubtedly cooler than the atmosphere, so the refraction could be enhanced under such conditions by a colder, denser skin of air in contact with the mountain – a tiny bit of anomalous refraction.

The reason for the elusiveness of the green flash is not absolutely certain, but we do know some effects that tend to reduce its visibility. The first and most obvious is that the sunlight is so reddened that no blue or green light remains to be seen. Countering that reason is the fact that we saw a most beautiful green flash one evening in Hawaii. We were seated only 2 m (6 ft) above the surf in the sea patio of the Halikailani Hotel. The sun set to a sea horizon below some cumulus clouds and into considerable humidity haze. The sun was rather reddened, as was the adjacent sky. We watched for the green flash, remarking that these certainly were not good circumstances. To our surprise, the moment the sun set a glowing green segment remained for about a second. It looked unusually wide (high), but we were so startled that our powers of perception could not be trusted. We did not have a second opportunity, as we continued on our way westward over the Pacific the next day.

Certain subtle effects need to be considered in seeking an explanation for the green flash. In addition to refraction we need to know the spectral sensitivity of the eye and the distribution of energy by color for the sun. Beyond this we need to consider the physiological reaction of the observer when the scene is flooded by colored light, as at sunset. First, let us look at the eye.

Analysis of the variation of visual response of the eye (Figure 3-2)

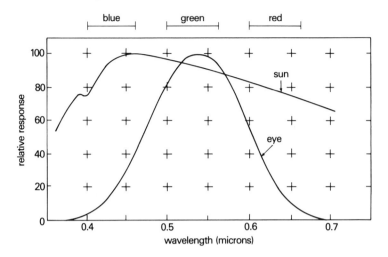

Figure 3-2. Variation of the sensitivity of the eye to color compared with the distribution of intensity of the sun, by color.

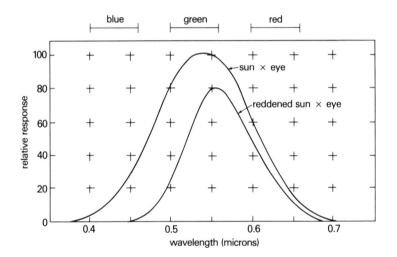

Figure 3-3. Variation of the color sensitivity of the eye combined with sunlight. Even reddened sunlight still has considerable green content to contribute to the green-flash effect.

shows that the eye is especially sensitive to green light, and much less so to blue or red. The spectrum of the sun also peaks in the green, slightly to the blue side of the peak sensitivity of the eye. The important factor is the net response of the eye to sunlight, especially to reddened sunlight.

Atmospheric absorption is highest in the blue, less in the green, and still less in the red. In addition, some bands of water-vapor absorption cut out sunlight in the yellow and red. Figure 3-3 shows two curves, one for the unreddened sun plus the eye and a second for reddened sun plus the eye. Note that when atmospheric absorption has been enough to eliminate the blue light from the sun the peak of sensitivity only shifts from the green to the yellow-green. Considering the effect of atmospheric refraction in cutting off the red image of the sun, we would expect the green flash to be more of a yellow-green flash. Yet to the eye it can appear emerald green, as we know from testimony of others and the truth of our own eyes.

Then can there be a blue flash? Yes. If the atmosphere is exceedingly clear (as it can be at Mont Blanc) your eye responds to the remaining visual signal of "blue," as did Lord Kelvin's.

Our search for pictures of the green flash shows how difficult that glint is to capture. We had been offered copies of the originals taken by O'Connell and Treusch at the Vatican Observatory at Castel Gandolfo, but these were taken through a telescope for scientific purposes and are not representative of what you will see with the unaided eye. Your eye does not see the magnified image shown through a telescope. The green flash occurs for only the last tiny segment of the solar disk. *Sky and Telescope* is a magazine that frequently publishes interesting pictures contributed by skilled observers from around the world. Looking through back issues, we soon found a number of pictures of the green flash in which the angular size of the sun is close to what you would see. One, the sequence of several camera frames in Plate 3-2, shows a pale green tint. A second sequence, taken from a movie of the setting sun and shown in Plate 3-3, has a bit more color change. The two frames in Plate 3-4 have a larger scale than you would see with your unaided eye, and the color change is quite noticeable. Plate 3-5 shows five frames taken through a telescope at the Vatican Observatory. You can now see that the greenish image of the sun is actually above the water horizon, an indication that some anomalous refraction must be present.

One quickly comes to the conclusion that the visual impact of an emerald-green flash is not the same as is recorded by the camera. What causes this disparity? Note that in all cases the sky background is heavily reddened. When you look at these photographs you see only a small piece of orange sky and the tiny image of the limb of the sun. If you were standing at the ocean your entire view would be filled with this deep coloring. It is well known that the human eye adapts to an overall colored scene, canceling to a considerable degree the actual hue and substituting one closer to expectation. Because of this effect Eastman Kodak suggests not photographing near sunset. If you do, the entire scene will be yellowed or reddened, even though your eye perceived the scene to be in daylight colors.

The experimental verification of this property of the eye by Edwin Land of Polaroid fame is dramatic. He showed that one can evoke the sensation of normal colors when in fact the total range of color in a scene is only from red to yellow. One can see green and blue even when in reality they are absent. The key is that the eye is deprived of seeing any true green or blue color. This is close to the situation for the green flash. The eye is looking intently toward a distant horizon where only reds and yellows prevail. A shift toward the green by refraction is accentuated by the eye and brain into a more vivid greenish tinge to the disappearing solar image. A yellow-green flash thus becomes an emerald-green flash.

The fullest explanation of the true origin of the green flash was made possible by the research of Father O'Connell aided by his associate at the Vatican Observatory, Father C. Treusch, who did the photography reproduced in Plate 3-5. Father O'Connell relates how, after returning to Italy from his unsuccessful attempts to observe the green flash at Lick Observatory in California, he was sitting at his desk at Castel Gandolfo in the Alban Hills southwest of Rome when he was startled to see a fine green flash as the sun set over the Mediterranean. He decided the observatory's telescopes were perfectly situated for the study of the phenomenon.[4]

One of the best clues that anomalous refraction is the chief culprit in the capriciousness of the green flash is shown in Plate 3-5. In the middle exposure we see green at each end of the segments, but red above, over the center. Red above means the reverse of normal dispersion. The second clue is in the lower exposure where the green flash is seen as a band of reasonable width separated clearly above the sea horizon. The cause is anomalous refraction of the atmosphere. If strong layering does not exist, the refraction will be "normal" and no green flash will be seen.

In the atmosphere at times temperature and humidity do not decrease uniformly with increasing height. The atmosphere is often layered, with rather different temperatures between adjacent layers. This interface between layers can also be wavy or turbulent, increasing or decreasing normal refraction. The sea–air interface is particularly subject to large temperature and humidity differences; thus, anomalous refraction effects are frequently seen over oceans at certain times of year.

This nonhomogeneity of the atmosphere is especially noticeable in the Chinese-lantern effect discussed in Chapter 2; thus one would expect to see evidence for this effect on evenings when there is a good probability for seeing a green flash. Passage of the upper or lower edge of the sun through a layer can momentarily amplify color effects of refraction.

Getting good pictures of the green flash is a challenge we hope some readers will accept. Timing is critical; so is exposure. Overexposure quickly dilutes the color intensity in color films. A good picture of the green flash will essentially silhouette the rest of the scene because all except the solar disk will be much underexposed. Good luck.

If you want the best chance of seeing and photographing this elusive phenomenon, we recommend a 1,000-mm-focal-length Celestron C-90 telescope or its equivalent. Attach it to your camera and use it as a finder. A binocular is dangerous for looking at the sun, even a sun heavily attenuated by haze, but the viewfinder renders sun viewing safer. The 1,000-mm scale makes the disk of the sun large enough to visualize the green rim shown in the Vatican photographs. The most striking effect, impossible to show in any photograph, is the dynamic progression of views. We had a chance to photograph the sun setting over the Pacific Ocean from La Jolla, California, but it was setting into a distant fog bank rather than the sea horizon. As the irregular cloud tops and the rim of the sun touched, green and blue colors danced and flickered here and there along the solar limb so startling us that we stood entranced, failing to press the shutter release until too late. Perhaps you can have the same thrill.

4

The earth's shadow
and sunset phenomena

We term "normal," sunsets that have not been augmented by transient injections of volcanic or meteoric dust. In reality there is only a subtle gradation between "normal" and "enhanced," depending on the amount of scattering particles in the upper atmosphere. Every year there are volcanic eruptions around the world, so there is always some volcanic aerosol aloft. Only an expert can detect the subtle differences between years until some really large eruption occurs.

To appreciate volcanic sunsets (Chapters 5 through 8), we must begin with normal sunsets. Let us first look at the sequence of scenes on a clear evening. The sun sets and the sky slowly darkens, with pale colors in the west and fainter colors in the east. The train of events is produced by the sun's dropping to a greater and greater angle below the horizon.

EARTH'S SHADOW

Within a few minutes of sunset, the edge of the earth's shadow appears above the eastern horizon opposite the sun, as shown in Plate 4-1. The greater the amount of haze in the atmosphere and the brighter the sunlight passing the horizon tangent point, the more visible the earth's shadow will be. The shadow is bluish-gray and shows the earth's curvature quite distinctly. Just above the edge of the shadow the whitish-blue may be tinged with a faint red tint, as in Plates 4-1 and 4-2. In Plate 4-1, taken at our home, the shadow is just rising and the full moon is above the shadow. In Plate 4-2, taken on Kitt Peak, the shadow is at its most striking.

The explanation of the curved edge of the earth's shadow and its coloration is simple. Consider the diagram in Figure 4-1. A zone of atmospheric haze is shown under the dashed line. The sky is clear to

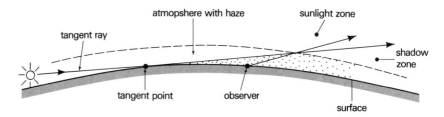

Figure 4-1. Geometry leading to the formation of the earth's shadow cast in the eastern sky shortly after sunset. A moderate amount of atmospheric haze enhances the effect.

the west beyond the observer's horizon, so sunlight can pass close to the earth's surface, grazing it at the tangent point. When the observer looks east, a portion of the haze layer is in darkness, but at an angle above the eastern horizon the haze is in direct sunlight, sunlight that has been reddened by its passage through the lower atmosphere at the tangent point.

The haze in the air scatters sunlight. Most is forward-scattered, which is why twilight colors are much stronger at the western horizon than at the eastern; but some is back-scattered. The red coloration is simply produced by the passage of sunlight close to the earth at the tangent point. An observer at the tangent point would see the reddened sun about to set, while the observer in the shadow zone would see evidence of the reddened sunlight only as that scattered by the atmosphere above him.

If it is cloudy at the tangent point, the clouds screen off the rays from passing close to the surface. In this case, the solar rays are scarcely absorbed and there is little color. On the other hand, the sunlight above the screening clouds is stronger and a very conspicuous earth's shadow may be seen.

High altitude also enhances the earth's shadow because the observer is looking edge-on at the haze layer and the shadow line. Plate 4-3 is a photograph of such a case that we took from a jet flying over Saudi Arabia. There was considerable desert dust in the air, reaching even higher than our plane, which about 30 minutes earlier had left Cairo. The full moon was above the shadow, making quite a sight. One can often see excellent views of the earth's shadow while flying at jet altitudes.

The edge of the earth's shadow appears to rise rapidly, somewhat faster than the sun sets, but it also becomes less distinct, disappearing when it reaches 10° to 15° above the eastern horizon. It is less visible simply because fewer air masses are in the line of sight, and hence there is less atmosphere to backscatter the sunlight.

LUNAR ECLIPSES

The earth's shadow can also be seen at the time of a lunar eclipse, cast on the full moon. Some of the same colors can be seen, mainly the reddish light from the setting sun bent by atmospheric refraction so that it fills in the shadow of the earth. The usual lunar eclipse thus shows a coppery red illumination of the moon. At times when the atmosphere is especially cloudy or when the high atmosphere contains volcanic dust, this reddish light is extinguished and the lunar eclipse is "black." Such black eclipses occurred after the eruptions of Krakatoa in 1883, Agung in 1963, and El Chichón in 1982.

MOUNTAIN SHADOWS

If you are on a mountain top something interesting is added to the earth's shadow. At sunset the shadow of the mountain can be seen cast onto the atmosphere to the east. Inasmuch as the sun has already set for the terrain in the valley, the earth's shadow will also be seen in the eastern sky. The shadow of the mountain is not in the shape of the mountain, but is triangular. Examples of this effect are shown in Plates 4-4 and 4-5, which are photographs of the shadow of Kitt Peak taken from two different locations on the summit. To obtain Plate 4-5, we stood exactly at the base of the 4-m (157-in.) telescope, and the shadow of this tall building can be seen at the peak of the mountain shadow.

The reason the triangular shadow converges to the edge of the earth's shadow is one of geometry. The shadows of things to the north or south of the observer appear to converge at the antisolar point, as do crepuscular rays (discussed later in this chapter). The light scattered back to the observer on the peak originates from the entire volume of scattering material in the air to the east. Imagine this material to be thin, vertical sheets situated some distance apart. Each acts as a gauze-thin projection screen and has the precise shape of the mountain silhouetted upon it. The shadows on the farther screen appear to be progressively smaller, but to an observer standing on the peak the points of the peak's shadow always lie superimposed because that is where the observer is located.

This situation is illustrated in Figure 4-2, where the observer is atop a telescope building. If we mark points A, B, C on the mountain, their respective points on each shadow screen, A′, A″, A‴, and so forth, lie on straight lines, all converging to the shadow point of the observer. The resulting shadow thus appears as a triangle with a vertex that

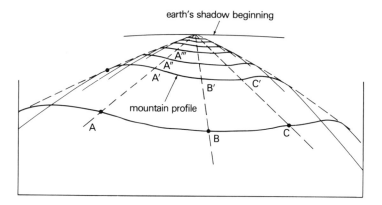

Figure 4-2. Geometry showing the development of the shape of the shadow cone from a mountain profile.

relates mainly to the height of the observer above the mountain top and only little to the shape of the mountain. Even if the mountain were a flat-topped mesa, the shadow would still come to a conspicuous point, emphasized because the total path length inside the shadow is at the vanishing point, the antisolar point.

The transitory nature of the mountain shadow was clearly described by Charles F. Lummis during his trek to California in 1884. While standing on the summit of Pike's Peak,

> with the setting of the sun came a sight even more memorable. As the red disk sank behind the west, the gigantic shadow of the peak crept up on the foothills, lept across to the plains, and climbed at last the far horizon and stood high in the paling heavens, a vast, shadowy pyramid. It is a startling thing to see a shadow in the sky. For a few moments it lingers and then fades in the slow twilight.[1]

It could indeed have been an exceptional sight because 1884 was the year after the eruption of Krakatoa had put much volcanic debris high in the atmosphere, providing a screen on which the mountain shadow was projected (see Chapters 5 through 7).

CLOUD RAYS

In summer one often sees cloud rays: blue-gray shafts of shadow. They can be seen at any time of day if the atmosphere below the clouds is hazy. Most often the good views of cloud rays occur near sunset. In the peaceful Hawaiian scene of Plate 4-6, the rays are beginning to form while the sun is still well above the horizon. Sometimes the rays are colored, as in the spectacular display of sunset colors during a summer storm in Tucson, shown in Plate 4-7, but more often they are bluish, as in Plate 4-8. The striking pink rays in Plate 4-7 are visible because the reddened sunlight is scattered by the haze and seen against the dark cloud background.

As time of sunset approaches the cloud shadows become almost parallel with the horizon, stretching far to the east. In Plate 4-9 we see such shadow rays, with sunlight still on the tops of relatively nearby clouds. Sometimes these rays can appear in a cloudless sky because the clouds causing the shadows are below the horizon. These rays mark the paths of beams of sunlight scattered by the dust and aerosols in the atmosphere. They thus are most conspicuous during times of air stagnation in late summer, when both natural and man-made aerosols conspire to give us hazy days.

CREPUSCULAR (SHADOW) RAYS

About the time of sunset on a hazy Indian-summer evening one can occasionally see faint bluish fingers stretching across the sky, converging in the east at the antisolar point. These are crepuscular rays. In Asia they are called the rays of Buddha, a term that appears also to include the bright shafts of light often seen in the western sky at the same time. The bright-colored rays are the shafts of sunlight scattered by the hazy atmosphere, and the bluish rays are shadows. Their origin is clouds, usually cumulus clouds of late summer, rising in a clear sky to the west of the observer. If these clouds are visible above your local horizon, you will see that their summits are still illuminated by sunlight. The most spectacular displays, however, are when the sky is perfectly clear; the clouds causing the rays are below the horizon. A nice occasion of crepuscular rays is shown in Plate 4-10, seen against a colorful clear-sky sunset. We saw this display from Chungli, Taiwan, and at first thought it was a volcanic sunset of the type described in Chapter 6; but the rays disappeared rather soon after sunset, indicating a low altitude for the dust producing the glow. Further evidence for a low altitude is shown by the convergence of the rays not far below the horizon. This point is the distance that the sun is below the horizon. The strong coloration in this case was probably due to fine dust carried out over the Pacific from Mongolia. Winds frequently carry desert loess over China, and that year loess was reported as far west as Hawaii. Desert dust often is carried far: In 1982 Florida had a strong incursion of dust from the Sahara.

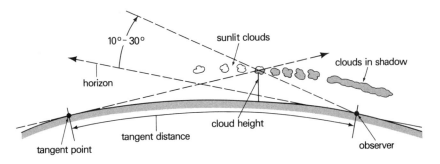

Figure 4-3. Geometry of sunset illumination on clouds.

Crepuscular rays sometimes can be seen stretching across the entire sky. Two such occasions are shown in Plates 4-11 and 4-12 taken from Tucson, and Plate 4-13 taken from Venezuela. The ray of Plate 4-11 is shown crossing the sky in the wide-angle photograph of Plate 4-12. When multiple rays converge to the antisolar point in the east, one obtains the display shown in Plate 4-13.

NORMAL SUNSET COLORATION

Plates 4-14 to 4-16 show a sequence of pictures of a normal sunset. In Plate 4-14 the lower clouds are illuminated with reddened sunlight. In Plate 4-15 the lower clouds are in darkness and the high cirrus are in reddened sunlight, yielding a nice pink because these clouds are very thin and some of the blue of the higher atmosphere comes through the cloud. This thinness is quite evident in Plate 4-16, where the cirrus no longer is seen, being also in the earth's shadow. The blue of the daylight sky is now turning a purplish hue.

The geometry of such a sunset is illustrated in Figure 4-3. The length of time a cloud is illuminated depends on the cloud height. Clouds generally lie in the range of 3 to 10 km (10,000 to 35,000 ft); direct sun illumination on the highest of such clouds lasts only about 17 minutes after sunset for an observer at 40° latitude. It lasts longer at higher latitudes; shorter at lower.

For a good color display at sunset the sunlight must come from below, as illustrated in Figure 4-3. Sometimes the sun can be seen setting below clouds, but no nice coloration on the clouds follows. This is because the clear spaces between clouds are not large enough for the sunlight to reach through to the clouds.

At sunrise the reverse sequence of events occurs. At some places the sunset sky is more beautiful than the sunrise, and at other places the reverse. This depends on where clouds tend to lie. In the far western United States the skies to the west, over the ocean, tend to be clearer than over the mountains to the east, making sunsets more colorful than sunrises. In any location the local weather pattern has much to do with whether the sunset or sunrise is brilliant. In urban surroundings one cannot enjoy the clear-sky sunset to the same extent, but sunset can subtly color even local smog.

SECONDARY SUNSET COLORATION

Under some conditions, especially after a brilliant sunset has dropped below the horizon, a strong secondary coloration can occur. The source of illumination of nearby clouds is softer; hence, the color is a blend of all the colors in the primary sunset. The secondary sunset is fainter but conspicuous on such occasions because the background sky is darker. The secondary glow does not appear until about 20 to 25 minutes after the sun has set, again depending on the altitude of the clouds involved.

CLEAR-SKY SUNSET COLORATION

At the time of sunset on a clear day the western horizon is a light yellow, shading up into the daytime blue of the sky. When the earth's shadow has crossed the zenith and is in the western sky, a faint straw or pinkish tint can often be seen in a circular "shield" reaching 20° to 30° above the position of the sun below the horizon, as shown in Plate 4-17.

The coloration of the setting sun and the twilight glow can be strong enough to tint delicately a snow-covered landscape. The setting sun often shines strongly enough over the cloud-free areas of France thus to tint the snows and clouds on the peaks of the Alps, producing the famous Alpine glow.

As the sun drops lower below the horizon, the line of demarcation between twilight and night sinks in the west, the sulfurous green squeezing the golden glow toward the horizon. In the west, if it is clear beyond the horizon, a reddish tint can be seen near the boundary of light and shadow. The brightest stars become visible overhead and in the east. If Venus or the thin crescent moon can be seen above the western horizon, the scene is one of surpassing quiet beauty.

PURPLE LIGHT

A pale purple or lilac glow occasionally develops at the beginning of a normal sunset when atmospheric conditions are favorable. This occurs when it is clear to the west at the tangent point so that the solar rays grazing the earth's surface are reddened but not extinguished by local haze. These reddened rays then continue upward through the atmosphere closer to the observer. If this local atmosphere is hazy some of the red light is scattered down to the observer. At just the right time after sunset there is still enough blue light coming from the sunlit zone in the atmosphere to mix with the red from the lower fringe of the sunlit zone to produce the somewhat rare purple light. An excellent example of this purple or lilac glow is shown in Plate 4-18.

To see the purple light in a clear sunset sky, look about 20° to 30° above the western horizon about 12 minutes after sunset. In recent years occasions of spectacular purple-light glows have been rather frequent, but these enhanced glows are due to unusual conditions in the high atmosphere that we will address in the following chapters.

TWILIGHT TO NIGHT

Scientific observation of precisely how the sky darkens after sunset was useful in deducing the nature of the atmosphere up to altitudes as high as 50 km (30 mi) long before direct sampling was possible. The book *Twilight* by G. V. Rozenburg details and summarizes the extensive researches of this type.[2]

How long does twilight last after sunset? The decrease in brightness of the zenith sky from noon to the onset of astronomical night is very great. The range of brightness with solar altitude is shown in Figure 4-4. It is too great to be shown adequately in a single graph, so we have provided a main graph and an inset with a magnification of 100. Civil nightfall is when outside activities cease in the absence of artificial lights. It is defined as being when the sun is 6° below the horizon.

Plate 4-19 shows the sky as it appears in deepening twilight, dark enough for stars to begin to appear. The sun still shines on the highest reach of the atmosphere, as shown by the fluorescence of a trail of chemical vapor injected over Yuma, Arizona, by the high-altitude research projectile.

As explained in Chapter 2, the eye is a logarithmic detector; hence, the sense of brightness change is not as dramatic as the reduction of brightness shown in Figure 4-4 of about 400,000 times from that at

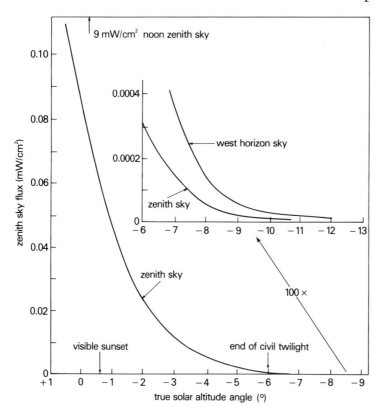

Figure 4-4. Decrease of sky brightness after sunset. The range is so great on a linear brightness scale that it is not feasible to show the brightness beyond a solar altitude angle of −13°. The brightness of the noon sky is in the order of 9 mW/cm².

sunset, or 70 million times from the brightness of the noon sky. For this reason the brightness of stars has from ancient times been expressed by magnitude, which is a logarithmic scale. The definition of the modern astronomical magnitude is that an intensity factor of 100 is 5 magnitudes. The faintest star you can see is 6th magnitude and the brightest star, Sirius, is −1.4. The difference, 7.4 magnitudes, is a difference in brightness of 912 times, yet the eye comfortably sees the faint stars in the constellation Canis Major without being blinded by Sirius.

If we plot the change in twilight-sky brightness on a logarithmic scale, as in Figure 4-5, we see that in the absence of artificial lights the brightness decreases almost linearly until the sun reaches −11°, whereupon it is slowly lost against the natural illumination of the night sky. This "light of the night sky" is caused by the vast number of stars too faint to be seen individually and by the airglow originating 60 to 200

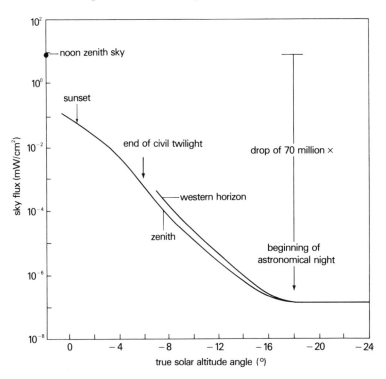

Figure 4-5. Decrease of twilight-sky brightness after sunset on a logarithmic scale, approximating the eye response, of about 70 million times the brightness at noon.

km (37 to 124 mi) above the earth's surface from chemical reactions, an echo of the absorption of sunlight by the high atmosphere. There also is a faint component from interplanetary dust scattered throughout the solar system, an extension over the sky of the zodiacal light. Astronomical night is generally assumed to start at a sun angle of −18°. when there is no longer any sunlight component in the western sky.

5
Volcanic eruptions

In Chapter 4 we discussed the appearance of sunset in a normal clear atmosphere. For years up to 1963 we enjoyed normal sunsets. Then something happened that, with subsequent events, has led to a significant change in the appearances of sunsets. On 17 March 1963 Mount Agung on the island of Bali erupted. Newspapers reported the event, remarking that several thousand people living on the slopes of the volcano had perished. Nothing more was reported, and we forgot about the eruption.

In September 1963 we became aware of some beautiful twilight glows, but thought little about their cause. Near the end of September Bart Bok flew from Canberra, Australia, to Tucson to succeed Aden as director of the University of Arizona's Steward Observatory. He mentioned hazy skies above his high-altitude flight and remarked that the haze had been over Australia ever since Agung had erupted. Again, we failed to make the connection.

Traveling a few days later to McDonald Observatory near Fort Davis, Texas, we noticed a brilliant orange skyglow through broken clouds when it was already dark above us. We remembered reading years before in geology class that the great eruption of Krakatoa had produced spectacular "sunsets" around the world. Suddenly we thought, Could this be from Agung?

The skies were clear upon our return to Tucson, and we noted that each night the brilliant twilight arch of golden light set 45 minutes after sunset. We immediately recognized that the glow was coming from a very high altitude. Some quick calculations based on the geometry of sunset showed that the glow was at an altitude of 18 to 20 km (65,000 ft), far above where clouds are found. Armed only with the name, Krakatoa, and the year, 1883, we headed for the university's science library.

Our first search opened a gold mine in the British science journal *Nature* and the American journal *Science,* then newly organized by Thomas Edison. One of the first references to Krakatoa was a notice from the log of the steam barge *William H. Besse,* en route from Batavia (Jakarta) for England. Much shipping from the Far East to Europe had to pass through the Sunda Strait, between Java and Sumatra. Astride this narrow neck of ocean lay the volcano-island of Krakatoa. It had been noted as "working" for months before the *William H. Besse* approached it. The sight that was beheld was "one of the wildest and most awful scenes imaginable."[1]

Many issues of both journals after the eruption carried articles that painted a drama of the worldwide impact of this eruption. One report noted that the barometers in England recorded the passage of an atmospheric pressure wave from the eruption of Krakatoa not once but on several successive passages around the world. Sunset phenomena were reported from many countries. The resulting tidal wave was noted on tide gauges around the world.

Curious to know more about volcanoes, we investigated the library's book section on geology, where Marjorie found a large, thick volume bound in aged leather: *The eruption of Krakatoa.* Opening it to the title page, she saw the added title, *"and subsequent phenomena."* This was the 1888 report of the Krakatoa Committee of the Royal Society of London. Scanning the table of contents we came to a chapter entitled "Height of the glow stratum."

Our immediate reaction was, What we are seeing *is* a glow stratum! But what about the altitude? Turning to the chapter, we found there were many height observations, averaging 66,000 ft (20 km)! We were observing the same thing seen almost a century earlier by another generation of scientists, amateur scientists, and laymen.

Excited by our discovery, we proceeded to look for descriptions of other eruptions to see what effects they had caused. The results yielded a puzzle. Some eruptions caused hazy skies and twilight glows. Others did not. Why? Further, Krakatoa had been an immense eruption, but Agung was small in comparison. Why had Agung injected material to the same very high altitude as had Krakatoa? Then 8 months later Surtsey began erupting, rising out of the sea southwest of Iceland. That eruption was small but long-lasting. Did it enhance the sky glows still existing from Agung? More recently, what about the violent eruption of Mount Saint Helens in Washington in May 1980? We will answer these questions in succeeding chapters.

TYPES OF ERUPTIONS

Geologists recognize several types of volcanoes based on the type of eruption and the geologic situation causing them. The gentlest are the shield volcanoes, typified by those in Hawaii. Their eruptions are relatively quiet, the most destructive action being the lava flows from the flanks of the mountain. In shape the shield volcanoes are broad, flattened domes. They produce small eruptions frequently. Mount Etna in Sicily and Hekla in Iceland are of this lava-flow type.

More violent are the cone-shaped volcanoes. They are active less often than the shield volcanoes and put out mainly ash and cinders. Their lava flows, when they do occur, are generally small and often are only intrusive into dikes within the cone itself. Mount Fuji in Japan is typical of this class, as is Mount Saint Helens. Mountain-chain volcanoes and island-arc volcanoes usually are the cone-building type. When they erupt, the result can be a major destructive event. We say "can be" because one cannot predict even days before the eruption whether it will be a building eruption or a destructive one. Most eruptions are small, adding to the growing height of the cone. Take Mount Saint Helens as an example. Small eruptions, occurring 50 to 100 years apart, built a symmetrical cone, one of the most beautiful in the Cascade Range. But on 18 May 1980 the mountain was devastated by a destructive eruption that tore 460 m (1,500 ft) from its summit and burst out its north flank. Mount Bezymia on the Kamchatka Peninsula of Siberia erupted in 1956 in a similar flank eruption, producing a shape very similar to that of Mount Saint Helens. So did Mount Katmai in the Aleutians in 1912. These eruptions inject into the air vast quantities of lava in the form of fine ash that blankets the surrounding country for a distance of hundreds of kilometers. The finer ash and gases move up into the high atmosphere, eventually to encircle the earth.

Hawaii-type volcanoes are relatively safe. People live near such volcanoes, losing their property but seldom their lives to the slowly advancing lava. After an eruption people move back, knowing that another flow in that exact area is unlikely for a long time. Thus the flanks of Mount Etna hold many villages, towns, and the major metropolis of Catania in relative safety.

The Vesuvius type of volcano seems to lie between these two extremes. It builds a cone slowly between major eruptions, but has fluid lava flows. Upon occasion, as in A.D. 79, it can explode, burying cities in ash and cinders.

In the most extreme type of eruption the entire top of the mountain is reduced to ash: The eruptions of Mount Mazama in Oregon, which formed Crater Lake, and of Santorin, in the Aegean Sea, were of this type. Many volcanoes or remnants now inactive show the scars of such destructive terminal eruptions. The Valles Caldera in New Mexico is one example. Similar eruptions about that same time may have been what buried many prehistoric animals in Nebraska. Recently, bones of a cluster of camels were excavated from beneath the thick layer of ash that had buried them ages ago.

Volcanoes usually give signals when they are about to erupt. Flurries of small earthquakes are the usual precursors. If there are seismographs reasonably near, one can calculate the depth of the earthquake beneath the volcano. If the flurries are observed to come from lesser and lesser depths, that is a clear sign that magma is slowly working its way toward the surface. The pocket of magma is often charged with dissolved gases, mainly water vapor, and as in a bottle of soda water the gases are ready to foam and propel the molten rock. In the case of Mount Saint Helens the landslide of the entire north face of the mountain released pressure that allowed the dissolved gases instantly to convert the molten magma into a dense cloud of vaporized incandescent rock.

Most volcanoes advertise their nature by a crater or cone-shaped profile, and most show abundant signs of earlier eruptions. The recent eruption of El Chichón in southern Mexico was an exception. It was considered a site showing volcanic or related phenomena only as a hot spring – solfataric. The several villages in the region had no inkling of what was about to happen. The mountain was far from seismographical stations and if the earthquake flurries were noticed, they were hardly likely to cause much excitement, as every few years there are rather strong earthquakes along this seismically active belt in Mexico. When the first mighty eruption occurred in the spring of 1982, the death toll was great. Weather satellites recorded clearly the vastness of the eruption and the rapid spread of ash around the world between 5° and 30° N latitude, leading to sunset and twilight effects we describe in the following chapters.

In historic times there have been many major eruptions around the globe that produced worldwide effects. In early days when world communications were poor, the occurrence of such events is recorded only as "dry fogs" high in the atmosphere with attendant atmospheric scenes that terrified people. Now we have good communications and awareness, so a major eruption cannot escape notice. Some in the perpetually cloudy regions of the Aleutian island chain cannot be seen from ship or the few villages along the chain, but a big eruption cannot escape detection from the many airplanes that fly near this region.

KRAKATOA

The eruption of Krakatoa was unusually well observed because the mountain lay astride one of the major shipping lanes between Europe and the Far East. The island chain forming Sumatra, Java, and Bali contains many cone-type volcanoes – so many that a major eruption is likely every few decades. Many kill people and destroy villages, but the world takes little notice. Krakatoa was different. Although most Indonesian volcanoes are high, Krakatoa rose only 800 m (2,625 ft) above the sea. Krakatoa is actually a group of small islands lining the sunken walls of an ancient crater that had exploded in prehistoric times. Subsequent small eruptions then rebuilt the islands that existed in 1883.

On 20 May 1883 booming sounds were heard in the populous villages and towns lining the shores on either side of the Sunda Strait. On 21 May Krakatoa was observed in eruption, the ash and vapor cloud rising to 11 km (36,000 ft) altitude. Ashes began to fall on the towns. By early June the eruption had subsided, and the volcano was ignored by residents and ships because they had become accustomed to the continual groanings and belchings of Krakatoa. Some ships did explore the area and reported several vents had opened on Perboewatan, the smallest of the several peaks but the one centered on the ancient crater. The extensive forest covering of Krakatoa had perished, with only a few stark stumps marking the largest trees.

On the morning of 26 August Krakatoa stirred to new activity. Strong detonations were heard and the sky began to turn dark over Anjer and Telok Betong, two towns that were to feel the full impact of what was about to happen. Three ships in the process of passing through the Sunda Strait were to be eyewitnesses of the event. Here it is best to read the words of the captains of these ships, as published from reports sent to *Nature* and *Science* magazines.[2]

On 26 August the sailing vessel *Charles Bal,* only 30 km (20 mi) away, "was in a fearful position since 1700." The captain's log recorded, "at 1700 sky darkening, detonations stronger, pumice stones pouring down, rather big pieces, ashes, etc., continual. Terrible nights." Fortunately the *Charles Bal* was on the west side of Krakatoa, as close as 17 km (11 mi) at one point. Most of the eruption material headed west.

The ship *Berbice* was not so fortunate, being first about 60 km (40 mi) southwest. The *Berbice* reported:

At midnight ashes increased, pieces of pumice stones, thunder and lightning increased, fireballs fell on deck and were scattered about, fearful roaring, copper at the helm got hot; helmsman, captain, and several sailors were struck by electric discharges; sail over the hatches to prevent fire, helm tied, crew sent below, captain and master kept guard: 27th at 0200, all hands to shovel ash into the sea, about 3 ft [1 m] thick lying on deck.

At 2300 the *Charles Bal* was in a worse position, only 17 km (11 mi) southeast of Krakatoa.

Lightning continued. Lay by; Krakatoa visible in northwest, 11 miles distant; strong wind southwest, chains of fire appearing to descend and ascend between sky and the island, while on the southwest there seemed to be a continued roll of balls of white fire; the wind, though strong, was hot and choking, sulfurous, with a smell as of burning cinders, some pieces falling on us being like iron shot; and the lead from a bottom of 30 fathoms came up quite warm. From midnight to 0400 (27th) wind strong but very unsteady between SSW and WSW, impenetrable darkness continuing, the rumblings of Krakatoa less continuous, but more explosive with the sky one second intense blackness and the next a maze of fire; masthead and yardarms studded with corposants [Saint Elmo's fire] and a peculiarly pinkish flame coming from the clouds, which seemed to touch the mastheads and yardarms.

This transition from continuous to explosive behavior signaled the beginning of the tidal waves that in a few hours would culminate in a wave 115 m (377 ft) high that would kill 36,380 inhabitants of the shoreline villages. The erosion of the old island by the eruption was now apparently allowing sea water to pour into the crater, temporarily hiding the fiery pit. Soon renewed ejections served to deepen the crater, accentuating the erosion.

At 0600 the first large wave struck Anjer. One survivor noted:

I went out about 0515. After having talked with several persons, I saw the wave, still far off, rapidly making towards us. I ran away, was followed by the wave, fell down quite exhausted, but happily on a hill where the water could not reach me. Before my eyes all the houses along the beach were destroyed.

Another person wrote:

I was early at the beach. When I returned home I heard a cry, "The flood comes." On looking around I saw a high wave which I could not escape; I was lifted from the ground, but caught hold of a tree. Then I perceived several waves which followed the first; the place where Anjer had been before was covered by a turbulent sea, from which some trees and roofs of houses were peeping out. After the wave flowed off, I left the tree, and found myself in the midst of devastation.

Shortly after, the *Charles Bal* passed where the lighthouse had been and could get no answer. In fact, the crew could see no movement of any kind along the shores.

Another ship, the *Loudon,* some distance from the blackness of Krakatoa, reported at 0700:

An immense wave came on; the Loudon, under steam, turned her head to it, was lifted up, but kept well; now the wave rushed on to the beach, and before the eyes of the passengers and crew of the Loudon, houses disappeared; the Berouw [a ship that had been thrown up on the beach by the smaller waves of the evening of the 26th] was lifted up and carried over a mile [a few kilometers] into the land. The place where Telok Betong had been before was changed into a violent sea. Three other waves followed at short intervals.

Still another witness recalled:

I went to Kampong Kankong, about 1.4 km away, to see the destruction which the wave had caused the night before. After I was there I saw a wave rushing on to us; we hastened to the hills, the villagers following us. When I reached the barracks, I saw Kampong Kankong had disappeared, and so had other villages near the beach.

The drama, however, was far from over. At 1000 it became so dark for the *Loudon* that not even the outlines of the ship or the persons aboard were visible; the ship was stopped for the next 18 hours. The report in *Nature* continues:

Rain of mud covered the deck 0.5 meter thick. Needle of the compass violently agitated; barometer extremely high; breathing difficult through damp [sulfurous fumes]; some people got unwell and sleepy. After the darkness began [it was late forenoon] the sea became violent, the wind increased; at last it was a hurricane. Then several heavy seas came, some of which came across and almost capsized the vessel. The flash of lightning struck the Loudon seven times, went along the conductor, but, when still above the deck, sprang into the sea. This was accompanied by a dreadful crackling. At such moments the vessel and the surroundings were brightly lighted; it was a fearful sight.

At 1115, the *Charles Bal* reported:

There was a dreadful explosion in the direction of Krakatoa, now over 30 miles distant. We saw a wave rush right on to the Button island, apparently sweeping right over the south part, and rising half way up to the north and east sides. This we saw repeated twice, but the helmsman says he saw it once before. At the same time the sky rapidly covered in . . . by 1130 we were enclosed in a darkness that might also be felt, and at the same time commenced a downpour of mud and sand which put out the side lights. At noon the darkness was so intense that we had to grope our way about the decks, and although speaking to each other on the poop, yet could not see each other. This horrible state and downpour continued till 1330, the roarings of the volcano and lightnings being something fearful.

The *William Besse,* having left Batavia for Boston, headed for the Sunda Strait and while some distance away reported on 27 August:

At daylight noticed a heavy bank [of clouds] to the westward which continued to rise; and the sun becoming obscured, it commenced to grow dark. The barometer fell suddenly to 29.50 and suddenly rose to 30.60. Called all hands, furled everything securely and let out the port anchor with all the chain in the locker. By this time the squall struck us with terrific force and we let go the starboard anchor with 80 fathoms of chain. With the squall came a heavy shower of sand and ashes, and it had become by this time darker than the darkest night. The barometer continued to rise and fall an inch at a time. The wind was blowing a hurricane, but the water kept very smooth. A heavy rumbling, with reports like thunder was heard continually; the sky lit up with fork lightning running in all directions, while a strong smell of sulfur pervaded the air, making it difficult to breathe. Altogether, it formed one of the wildest and most awful scenes imaginable.

Aug. 28. Thick smoky weather, hove anchor.

Aug. 29. Passed Anjer, and could see no light in the lighthouse, and no signs of life on shore . . .

Aug 30. [just beyond the straits] Found the water for miles filled with large trees and driftwood, it being almost impossible to steer clear of them. Also passed large numbers of dead bodies and fish.

The *Berbice,* in the same area, reported: "When we passed Prince's Island we saw banks of pumice stone 18 to 24 inches thick. The sea was covered with pumice stones and floating corpses."

Following the mighty explosions on 27 August, the eruption of Krakatoa ceased completely, leaving only some steam arising from the cinders on the remaining parts of the island. Where the eruption had burst, the sea was now over 305 m (1,000 ft) deep. The sea had won this battle. Over the years since then, as a result of many minor erup-

46

tions, Anak Krakatau ("Son of Krakatoa") has appeared above the sea, building for a new battle some year in the next few centuries.

The eruption over, the world was destined to see phenomena in the sky for years after the last echo of the mighty blasts was gone.

CONSEQUENCES OF ERUPTIONS

Earlier in this chapter we mentioned finding a treasure on the Krakatoa phenomena in our university library, *The eruption of Krakatoa and subsequent phenomena,* compiled by the Royal Society of London. The events surrounding the Krakatoa eruption were so dramatic and so widely observed that they are worth looking at more closely before we proceed to a general discussion of volcanic sunsets and glows. We quote from the preface of the Royal Society report:[3]

The extremely violent nature of the eruption of Krakatoa on August 26th to 27th, 1883, was known in England very shortly after it occurred, but it was not until a month later that the exceptional character of some of the attendant phenomena was reported. Blue and green suns were stated to have been seen in various tropical countries: Then came records of a peculiar haze; in November the extraordinary twilight glows in the British Isles commanded general attention; and their probable connection with Krakatoa was pointed out by various writers.

At the 10 January 1884 meeting of the Royal Society, President Julian Huxley offered a resolution "that a committee . . . be appointed to collect the various accounts of the volcanic eruption of Krakatoa, and attendant phenomena, in such form as shall best provide for their presentation, and promote their usefulness."

This volume not only summarized the worldwide events that followed the eruption but also reviewed earlier records to see if any since 1500 had rivaled Krakatoa. The key words sought in old literature were unusual atmospheric phenomena such as blue suns, dry fogs, purple light, and red twilights. Quite often the records were from a single writer, inasmuch as few persons were attuned enough to nature to recognize something as unusual. We have found from our own experience that you become sensitive to differences in the twilight skies only when you are careful in regular observation. Our desert home overlooks the broad valley wherein Tucson lies, so that every evening from our home we see the sunset skies. Even so, it took the volcanic twilights after Agung's eruption to tune our sensitivity. We now know that the upper atmosphere is frequently augmented by eruptions, but with less dramatic results. Each year there are a number of volcanoes in various stages of eruption.

In the seventeenth century only the largest changes in the atmosphere were sufficient to gain notice and merit published report. Sometimes the consequences of an eruption were noted, but no specific reference to a large eruption was brought to the attention of scientists in Europe. In the list of eruptions we recognize volcanoes or regions where current eruptions have produced visible effects: Volcan de Fuego in Guatemala in 1581, 1799, and 1974; Mount Saint Augustine in Alaska in 1883 and 1976. Photographs of the sunset glows resulting from the recent eruptions are shown in Chapter 6.

The earliest records of volcanic twilights date from 1553, when remarkable purple glows were seen in Denmark, Sweden, and Norway. In 1636 a red glow was seen in the sky in Scandinavia, and arriving sailors reported that for weeks the sky had seemed on fire after sunset. A major eruption year was 1783 with Hekla (Iceland), Skaptar Jökull (Iceland), and Asama (Japan), the last being one of the most violent on record. The following year (1784) was unusually cold; in fact, the following decades marked what has become known as the Little Ice Age. These years were also a period of polar exploration, and the increased severity of the Arctic winters added hazard and hardship to the explorers. There are many references to the dry fogs and a sulfurous smell over Europe in the summer of 1784; the sun was heavily reddened, the sky blood red at the rising and setting of the sun.

A whole series of eruptions occurred between 1783 and 1818, several being very large: Saint Vincent in Martinique (1812), Mayon in the Philippines (1814), and Volcan de Fuego in Guatemala (1799); the last was followed by magnificent sunsets in an indigo sky.[4] Photographs of the sunsets from the 1974 eruption of Fuego are shown in Chapter 6. The eruption of Tambora in Java (1815) was an immense one, the largest since volcanic records have been kept. A lake now lies in its crater, blasted out from the original peak, a crater 6 km (4 mi) in diameter. Some 10,000 persons living in nearby villages were killed in the eruption and an additional 82,000 died subsequently from starvation and consequent disease. All these large eruptions produced brilliant twilight glows.

In 1831 there were large eruptions of Guagua Pichincha (Ecuador), of Ferdinandia (a temporary island also known as Graham's Island) in the Mediterranean between Sicily and Pantelleria, and of an unnamed volcano in the Babuyan Islands. Green and blue suns were seen along with spectacular sunsets and twilight glows. Large eruptions followed in 1835, 1837, 1845 (Hekla again), 1846, and 1855. In 1858 Cotopaxi in Ecuador sent a vast amount of dust into the sky, the fourth largest eruption after Tambora, Krakatoa, and the combined eruptions of Mount Pelée and Soufrière in the Antilles Islands in 1902. There was then a lull of 50 years until Agung erupted in 1963, save for Katmai in the Aleutians in 1912. Since 1963 several eruptions have enhanced the residual atmospheric effects of Agung: Awu in the Sangihe Islands, Irazú in Costa Rica, El Fuego in Guatemala, and most recently, El Chichón in Mexico. In fact, it has been hard to finish this book because of the many dramatic eruptions and their resulting effects on the high atmosphere.

Edward Wilson, an artist as well as an explorer and reporter, accompanied Sir Ernest Shackleton on his 1901 voyage to Antarctica and reported and painted many glorious sunrises and sunsets that sound and look much like ones caused by volcanic dust aloft. Wilson's note for Monday, 9 September, is especially interesting:

Shackle called me up about 5:30 A.M. to see a glorious sunrise. I tried the sketches but the immensity of this golden firmament is lost on paper . . . The sunset was even more glorious than the sunrise, for the sky was almost cloudless and we got the intense yellow ochreous glare after sunset uninterrupted by any clouds. It was almost uncanny. One felt as though something terrible was about to happen – the same sort of feeling that one gets in a dense yellow London fog, only this was beautiful, and magnificent, as well as terrifying. Everyone was on the bridge watching it.[5]

We find this description fits well with our impression of volcanic sunsets, especially after the 1982 eruption of El Chichón. The Agung sunsets had a clarity about them; the first El Chichón sunsets appeared ominous, the entire sunset sky looking as though some mighty hand was pulling a yellow-ochre roof across the sky. More of this in Chapter 6.

With this introduction to volcanic events we will proceed to look at what causes the atmospheric effects, recent photographs of these effects, and studies of current eruptions.

Not all violent eruptions produce large atmospheric effects. On the other hand, some lesser eruptions apparently do. The critical factor appears to be the amount of material injected into the atmosphere during the eruption. Some eruptions last only a few hours and are distributed as discrete explosive events of short duration. Vast destruction can be caused by such eruptions, as exemplified by Mount Saint Helens, which caused only slight atmospheric effects except immediately downwind of the event. A further criterion is the amount of sulfur dioxide gas emitted during the eruption. We will examine this in Chapter 6.

6

Volcanic twilights

It would be easy to continue a whole book about volcanic eruptions and their effects. We list in the Notes for Chapter 5 some references for those who are interested in reading more. Let us, however, resume our story of the Agung eruption sequence. Our early misconception about "volcanic sunsets" was that colored *clouds* made these sunsets spectacular; yet what we noticed was simply a brilliantly colored *glowing background*. We were excited when we turned the pages of the Royal Society's Krakatoa report to see a volcanic sunset referred to as a "glow stratum." In 1884 S. E. Bishop of Honolulu (a missionary-naturalist) described twilights surprisingly like what we had been seeing:

One marvelous effect is often a sudden appearance of thick luminous haze where a minute [*sic*] before all was pellucid, unsullied blue. Meantime the glow especially gathers and deepens above the western horizon along a line of 60 degrees until the whole occident is a uniform sheet of flaming crimson, shading up into lilac and orange. Down upon that creeps the dark earth-shadow, sharply cutting off the edge of the blazing sheet, often serrated with the shadow of remote cumuli. As the shadow descends, the glow deepens, until night has closed down upon it. At once on the darkened sky arises a secondary or "after-glow," repeating the same phenomenon as the stars come out with almost equal brilliancy of effect. In this after-glow the defined shadow line is lacking, and the deep fiery red above the horizon bears a singular resemblance to the peculiar reflection on the sky of some immense but remote conflagration. These appearances occur before sunrise with equal brilliancy, but in reversed order.[1]

We would add to the above description that the display usually begins about 15 minutes after sunset and that the horizon mountains and clouds are inky black silhouettes against the glowing stratum.

The description by Bishop is illustrated in Plates 6-1 to 6-3, taken from the summit of Kitt Peak National Observatory only a few days after we first noted the enhanced glows. As in Bishop's description there was a cloud shadow from a cloud below the horizon. A second set of photographs taken a few days later but in a much different climate, at Yerkes Observatory in Wisconsin, is shown in Plates 6-4 to 6-6. Both sets show the sequence beginning with the lilac atop orange, the glow deepening as the earth shadow creeps downward. The progression of the glow into the secondary glow is shown in Plates 6-7 to 6-9. In 6-7 the primary glow is already nearing the horizon beneath a local deck of clouds. The exposure time for each of these pictures is increased to compensate for the greatly diminished intensity of the secondary glow, already becoming apparent in Plate 6-8. The primary glow has set in Plate 6-9, and the city lights are visible, along with a thin crescent moon. To our eyes the crescent moon was greenish-blue, a fact not apparent in the photograph but explained by the shift in eye sensitivity discussed in Chapter 4.

One of the most fascinating things about the Agung sunsets was the variability of appearances. After the onset of magnificent glows, a photographer from *Time* magazine came to Kitt Peak expecting to take pictures like those in Plates 6-1 to 6-3. Instead, there was scarcely even a yellowing of the western horizon. He went away probably muttering to himself about overenthusiastic astronomers. He had seen much better sunsets in Chicago. At that point we were uncertain what had happened to the glows. Had the volcanic material gone away? The next day was cloudy, and we had the clue. Our weather comes from the west. Cloudy weather on the way screens the sunlight that would otherwise pass close to the earth and become reddened. If the reddened rays are screened out there is no color to the glow stratum.

Various mixtures of reddened sunlight and local sky color can lead to striking variations in the volcanic twilight glows. Plate 6-10 displays a lilac upper zone that verges on the best purple light possible. Even as seen from within a city the glows can be pictorial, as in Plate 6-11, taken near the maximum of the Agung glows. The widespread nature of the material from a major eruption means that very similar twilight glows are seen when the right weather conditions exist at quite different locations. Plate 6-12 is of fiery intensity to equal the national mourning that evening. It was taken in Boston on the evening of the funeral of President Kennedy.

Our observations of the time after sunset to the set of the glow on the horizon showed that the scattering layer was at an altitude of about

20 km (66,000 ft). The secondary glow, caused by the layer scattering the brilliance of the primary glow, now below the observer's horizon, persists 20 to 30 minutes after the primary has set. Like some of the observers quoted in the Krakatoa report, we at first interpreted this persisting secondary glow to scattering from a higher layer; but such a layer would have to be at about 46 km (150,000 ft), where there is not enough air to support anything of the nature of dust. We had leaned toward this explanation because sometimes the secondary glow was bright and sometimes missing, even though the primary was comparable in both cases. The correct answer became clear when we realized the connection with clouds to the far west. The primary glow seen from Tucson could be bright, but the primary glow below our horizon that was necessary to produce a good secondary could be faint because of a very distant cloud cover.

Shortly after we sighted the Agung sunsets, Aden had an opportunity to view some of the early films returned from space. Of particular interest were photographs of the horizon in which the stratification of the atmosphere was very distinct. Comparison of photos before and after Agung's eruption made it clear that a sharply defined stratum had formed high above the earth. Measurements gave a height of 18 km (59,000 ft) for the maximum haze and about 20 km (66,000 ft) for the top. This clearly was the stratum responsible for the twilight glows.

The role this layer plays in forming the visual color displays at twilight is clearly understood when we think of the layer as a movie projection screen held aloft. Sunlight from the set sun reaches it from below, and we too view it from below. The stratum "screen" scatters enough sunlight downward to make an upside-down picture of the shadow of the earth. The thicker the stratum, the brighter the display. Figure 6-1 shows this geometry. Of course, if the stratum is too dense – for instance close to the eruption – the brightness of sunlight is greatly diminished in passing the tangent point. A filmy stratum is necessary for best effects, so thin it is scarcely visible during daylight.

The visual appearance of the glowing twilight colors is explained by the diagram in Figure 6-2. The exact tints are a combined effect of the color of sunlight reaching the stratum and atmospheric filtering on the path from the stratum to the observer. The filtering of the atmosphere at the tangent point determines the colors reaching the stratum, and they can be quite different from those near the observer. Clouds below the horizon can create gaps in the solar illumination, creating blue rays that appear to diverge from the position of the sun below the horizon.

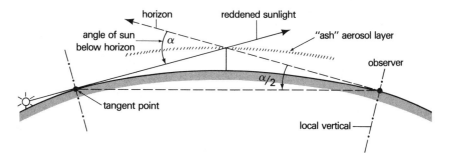

Figure 6-1. Basic geometry used to determine the height of clouds from the time of sunset on the cloud and the angle of the sun below the horizon.

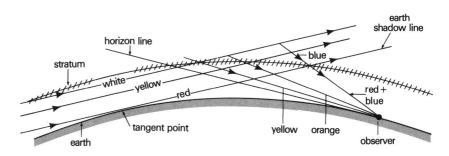

Figure 6-2. Origin of the sunset color sequence seen by the observer. The red and blue light combines to make the purple light.

The state flag of Arizona displays a gold-and-blue-rayed sunset, perhaps inspired by the volcanic sunset glows that occurred about the time Arizona adopted its flag.

The twilight glows that began with Agung continued for almost 6 years. During that time several other major eruptions occurred, probably adding to the scattering material from Agung. They were not as big as Agung, and we wondered why they should produce haze at the same altitude, and why, in fact, had Agung, a lesser eruption than Krakatoa, produced haze at the same altitude seen in 1883. We will answer this question in Chapter 7.

By 1974 we thought our enjoyment of volcanic twilights was only a memory not to be repeated in our lifetime. Our attention by then had turned to studying solar energy, though sunsets remained a continual enjoyment and were recorded in Marjorie's notebook. We were invited to go to Mexico City for the inaugural of the Mexican Academy of Science, to talk on the future promise of the Age of Solar Energy. En route we changed planes at Mazatlán, and took off from there just after sunset. Banking toward Mexico City, we saw the sunset sky through breaks in the local clouds. It was brilliant fiery orange!

We remarked to each other, Just like Agung! We saw no sunsets from Mexico City, but a week later while flying over Guadalajara on our return home, we looked westward and again saw a flaming sky. It must be a sign of a new eruption. We waited eagerly the next evening for what we would see from Tucson. Nothing.

But 6 days later, even before sunset, Aden looked out toward the west and saw a silvery stratum low on the horizon. "It's here," he shouted. We set up the camera and started a sequence. We would not know if the stratum was just a distant cloud layer, albeit an unusual color, until we saw when the sunlight ceased to illuminate it. The first picture, Plate 6-13, showed the ordinary yellowish color of any cirrus cloud shortly after sunset; we waited to see what would happen. As the twilight sky darkened and the cloud remained in sunlight, our excitement grew. But the next exposure (Plate 6-14) showed no lilac top – only the blue of a normal twilight. As the glow deepened to red (Plate 6-15), we knew by the time of glowset that we had seen in Tucson the same high-altitude glow we first saw a week earlier from Mexico.

The duration of the glow was 45 minutes, so it was a volcanic sunset. But where in the world was the eruption? Before going to Mexico we had read of an eruption of Fuego in Guatemala and thought little about it because that volcano is frequently active. Remembering we had seen the glow first from Mazatlán and from north of Guadalajara, we thought the eruption might have occurred from one of the volcanic islands off the tip of Baja California. Soon, however, letters arrived from friends telling of the violence of this eruption of Fuego.

The morning after we first saw the Fuego sunset in Tucson the entire sky was covered with silvery filaments of haze. Its appearance at the moment of sunset is shown in Plate 6-16. The brilliant colors typical of a volcanic glow stratum followed, muted because the thickness of the stratum must have screened the incoming sunlight. The sky's appearance 20 minutes after sunset is shown in Plate 6-17. The lilac color was back also. The glow deepened in intensity and reddened as twilight progressed (Plate 6-18). For about a week the glows were magnificent, but then dimmed almost to preeruption levels. About 2 weeks later the glows returned, though not as bright. We were elated: Even though it was fainter, this was our second discovery of a volcanic sunset! The eruption cloud from Fuego had slowly drifted westward over the Pacific and had been seen just east of Hawaii. It then had drifted eastward, crossing over the southwestern border of the United States – and over us. Since 1963, the year we first noted a volcanic sunset, a technique called lidar had been developed to detect atmospheric dust. Lidar consists of a short pulse of ruby laser light sent directly upward. Each dust layer sends a return echo, which is detected by an adjacent telescope and photomultiplier detector. The altitude of the scattering layer is given directly by the delay time. With this tool the Fuego dust was observed when it passed monitoring stations. But more about lidar in Chapter 7.

To illustrate the contrast between twilights with and without volcanic glows, we show two Christmas pictures: Plate 6-19, taken in 1968, and Plate 6-20, made in 1974. In 1968 only the faintest pink reminded us of the glory of the twilights enhanced by the eruption of Agung. In 1974 we were overwhelmed by the intensity of the Fuego glows. For comparison we show the sky's appearance when a distant cloud layer cuts off the upper portion of the glow (Plate 6-21), so that only a few fragments of pink light survive.

A year later, with the detection of two volcanic sunset events to our credit, we were astonished on Sunday afternoon, 25 January 1976, to look up at a clear blue Arizona sky and see soft, smoky silver windrows of cloud drifting in from the northwest. It looked like a volcanic ash cloud, but a third time? We took the photograph shown in Plate 6-22. Several hours later at sunset, the ashy cloud was still overhead, although the main part had rapidly passed to the southeast. The pale coloration (Plate 6-23) didn't look like a volcanic twilight. We calculated an altitude of only 12 km (39,000 ft), quite a bit lower than the altitude for the earlier glows. We consulted the bulletin on transient phenomena published by the Smithsonian Institution. There had been an eruption of Mount Saint Augustine, a known volcano in Alaska, a few days earlier. It seemed extraordinary that two ash clouds, one from an eruption far to the south in Guatemala and one from far to the north in Alaska could *both* happen to pass over Tucson. Because this new cloud was at a low enough altitude to be affected by winds, we examined meteorological maps to determine the air trajectory: There was a strong air flow aloft from Alaska. Calculating backward, we noted with satisfaction that the air mass containing our cloud had been, in fact, over Saint Augustine when the volcano erupted. We wrote of this to a friend in Fairbanks, Alaska, but his letter to us passed ours in the mail: Had we seen any evidence of the eruption cloud from Saint Augustine? We collaborated in writing a paper.[2]

It turned out that Mount Saint Augustine had produced only two rather small eruptions a few hours apart, with nothing significant before or after, so the aerosol layer showed no lasting effect. The cloud was visible only because the eruption was in a clear, cold mass of air, in which the eruption cloud was literally frozen, being still in the same very cold, clear air mass when it passed over Tucson. Later we saw some fine photographs of the eruption, taken from the NASA earth-imaging satellite, Landsat, and showing the ash being injected into the clear air. Frequently volcanoes that erupt along the Aleutian chain are hidden by the perpetual cloudiness of that region; only the top of an eruption cloud is seen by a passing airplane. Saint Augustine was not the end of our adventures with volcanic eruption clouds; however, it was quite unusual to have the clouds from a Central American eruption, El Fuego, and an Alaskan eruption, Saint Augustine, both enter the United States by passing over Tucson, Arizona.

When Mount Saint Helens exploded in 1980, it sent a dense cloud of ash over the northern United States and produced colored sunsets in northern latitudes, as we saw on a trip to Minneapolis. The sunsets over Tucson displayed a modest pink glow, shown with a cloud shadow that enhanced its visibility in Plate 6-24.

The Saint Helens glows rapidly diminished. On 24 August 1980, we noted in the west through foreground thunderheads some smoky wisps. A sunset glow developed on these wisps, ending about 45 minutes after sunset. More volcanic ash clouds? Again we investigated the air trajectory and identified the wisps as originating from the 17 August 1980 eruption of Hekla on Iceland. The glows continued for the next few days, but soon became faint. By January 1981 the sunset skies over Tucson had essentially dimmed to their pre-Agung level. For years after Agung, as the orange-red sunsets persisted, we had wondered if we would ever again see the blue twilight sky. By 1981 the blue twilight of a normal sunset had returned, but for how long?

Not long at all. We left in October 1980 for Taiwan, where we were to test the 60-cm (24-in.) telescope just completed for the new Institute for Physics and Astronomy at National Central University, a project we had helped start 2 years earlier. As soon as we arrived, we noticed how brilliant the sunsets were compared with what we had last seen from Tucson. One evening the rays were spectacular, as shown in Plate 4-10. Where did this material come from? We could not answer this question clearly, as there had been the usual number of small eruptions around the world in 1980. At the time we thought these rays were cloud shadows on some unknown volcanic aerosol, but we now are reasonably sure they were projected on a veil of high-altitude dust from Mongolia. Incursions of dust from the Mongolian deserts often drift westward over China, being detected on occasion even over

Hawaii. The Taiwan glows faded early, indicating a lower altitude than one would expect for a volcanic glow, though volcanic material sometimes occurs at a lower level than the aerosol layer normally is located.

As we were finishing this book some new aspects of volcanic sunsets were unfolding. We had read about a new eruption of the volcano El Chichón, also known as Chichonal, in Southeastern Mexico. It had destroyed several villages, with large loss of life. The site was not a known volcano, being classified as a solfatara, which means a noisy, sulfurous hot spring. The eruption was violent and continued for several months. Satellite photos showed that a vast dust cloud had been emitted, and some dust was seen as high as 25 to 30 km (82,000 to 98,000 ft). Our interest was roused when Brian Toon of NASA Ames Laboratory near San Francisco alerted us to be on the lookout for the cloud, which was reported over Hawaii and which might soon drift over the southern United States. Two days later, shortly after sunset we saw low in the west a silvery cloud bank much like the one from El Fuego. We told Toon; he dispatched a U-2 plane to gather samples that confirmed it was in fact the ash cloud from Chichonal. Even as we write this, the ash stratum is producing a continuing sunset display.

The Chichonal ash stratum is the heaviest and most persistent to date, so heavy it was visible all day long for about 2 weeks as a dull ashy veil, heavily rippled at times. The sun was surrounded by a strong aureole. The first aspect of sunset was direct unreddened sunlight, giving a silvery appearance to the ash, as shown in Plate 6-25. The progress of the twilight glow, however, was quite different from the development of the glows we had seen after earlier eruptions. The ash stratum was so heavy that it caused self-shadowing effects that changed the colors and dimmed them so much that almost no twilight glows were visible on the evenings of the heaviest ash clouds. As the sun neared the horizon on one of those occasions, long windrows of heavy ash stretched from southwest to northeast in a band 30 to 40 km (18 to 25 mi) wide and 200 to 300 km (125 to 190 mi) long. When the sun passed through these windrows they appeared a striking desiccated coppery color, with a vertical reflection bar above and below. We wondered about that bar, as its formation requires aligned flattened particles; similar bars form from ice crystals during winter sunsets. We read later that flattened particles had been gathered by sampling aircraft.

The extent of the self-shadowing is visible in the sequence depicted in Plates 6-26 to 6-29; the depth of obscuration is gradually diminished. In Plate 6-26 unreddened sunlight illuminates the ash with essentially no illumination above the silver layer. In Plate 6-27 the ash layer is patchy enough so that a few weak pinkish areas can be seen. The most ominous of the various appearances, occurring for about a week, developed where the layer was uniform and weakly passing sunlight of an ugly orange color above the whitish band (Plate 6-28). One had the feeling that the sky was falling as the whitish band narrowed and the ugly orange "roof" descended in the west. The effect of self-absorption slowly diminished and the shadow band was narrowed (Plate 6-29). This shadow band set ahead of the top of the glow, with fiery red showing as the glow set (Plate 6-30).

To gain a better understanding of the various appearances when the ash is heavy, examine Figure 6-3. The difference in the three cases shown is the angle of the sun below the horizon. In the top diagram the sun has just set and the sky is as shown at the left. Above the sun is a silvery shield with perhaps a touch of yellowish color at the edge of this aureole. All the solar rays shine on the ash layer *from above* and thus are not colored.

In the middle diagram the sun is below the horizon and the whitish band has shrunk toward the horizon, its color being due simply to passage of light from the ash layer through the lower atmosphere to the observer. Where the sunlight passes tangent to the ash layer (center right diagram), the long path of the rays through the ash greatly diminishes the sunlight, and the observer sees a shadow band separating the lower whitish band from the orange-pink upper glow, much as in Plate 6-29.

In the bottom diagram only the red remaining portion of the glow is above the horizon, and the last rays are tangent to the earth. It is these last rays that are cut off when a cloud layer exists at the tangent point.

Note that in all these appearances, the El Chichón glows were quite different from those of Agung and El Fuego in that lilac was missing. There apparently must be an ash layer of very tenuous nature to show the best colorations. Even as we write these comments the El Chichón glows are improving. Our most thrilling experience with volcanic glows occurred in late July 1982 when we saw a spectacular example over the Pacific Ocean at La Jolla, California. The sky was clear except for the usual evening fog patches forming out over the ocean. The primary glow that evening (Plate 6-31) featured a vivid orange sky and a deep orange ocean. The secondary glow, even more awesome, is shown

Sunsets, twilights, and evening skies

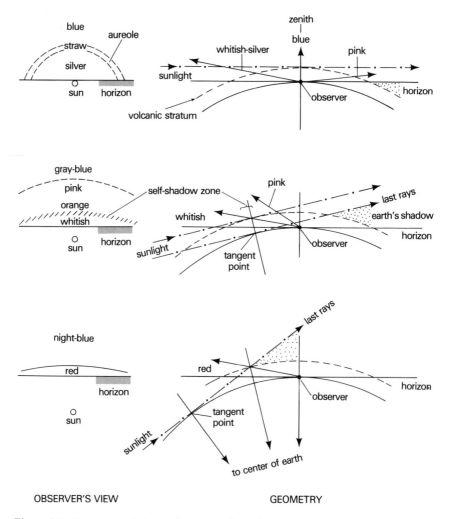

| OBSERVER'S VIEW | GEOMETRY |

Figure 6-3. Geometry and view of the western sky at three times after sunset. Top: At sunset with afternoon aureole of the sun still showing. Middle: About 20 minutes after sunset when the self-shadow zone of the volcanic-ash layer is seen 10° above the horizon, with unreddened sunlight shining down on the layer below the zone and with light dimmed and reddened by passage through the layer shining up on the layer above this zone. Bottom: About 40 minutes after sunset with only the last reddened rays shining upward onto the layer.

in Plate 6-32, made exactly 1 hour after sunset. The secondary developed unusually late; but the El Chichón ash passing overhead those days was unusually high, about 28 km (92,000 ft).

Another new insight into the secondary glow became apparent during the Chichonal sunsets: The ratio between primary and secondary glows is not the same for all events. We think the reason is that the sun, being effectively a point source, can be heavily diminished in intensity along the tangent path when the ash is heavy, but the integrated intensity of the primary glow is spread over a large solid angle and thus is collectively a bright illumination source for producing the secondary glow.

A great eruption of Galunggung in Indonesia in 1982 also promises to enhance sunsets along with Chichonal for some years to come. The ash ejected by this volcano has been very heavy, seriously damaging all four engines of a 747 airplane that strayed into the ash plume and giving passengers a strong whiff of volcanic gases. If past large volcanic events are a guide, you may still be seeing traces of the sunset glows until 1986 even if no further large eruptions occur. Thinking back to years before 1963, we remember no sunset like what we have seen in the 20 years since Agung. Is this only because our attention had not been sufficiently alerted to the differences between one sunset and another? We think not. Then does this mean that large volcanic eruptions producing stratospheric effects have suddenly become commonplace? If that is so, then what might be their accumulative effects on climate? Can enough science information be derived from these events to allow computers to predict what could happen if eruptions continue for a few more decades or a century? These and similar fascinating and important questions intrigue many young scientists who are pursuing volcanic events.

Seeing things in all creation is mostly being diligently aware of nature around you. Volcanoes distribute their ash and gases worldwide. You need only be alert and know what to look for so that you too can detect these changes in the sunset skies as the transition from day to night proceeds, summed with the appearance of the thin crescent moon in the twilight glow (Plate 6-33).

7

Twilight science

The brilliant sunset glows from the Agung eruption disappeared in the early spring of 1964, after half a year. We were not completely surprised; nor were the observers after Krakatoa when the same thing happened.[1] They had assumed the disappearance of the glows was an indication that the volcanic ash had settled out. They were startled to see the displays come back with "full intensity" the next summer. Their first thought was of some new eruption, but there had been none. Over the succeeding several years this phenomenon was repeated, while at the same time the glows gradually faded, so that after 6 years the sunsets had returned to their pre-Krakatoa level. Thus with some interest we awaited summer 1964. Would the Agung glows return? They did indeed. Marjorie has kept an accurate record of the brightness of the twilight displays continuously since Agung. The results for several years, until it became evident that new eruptions were enhancing the 20-km (66,000-ft) level, are shown in Figure 7-1.

VISUAL OBSERVATIONS

The visual estimates of the brightness of the clear-sky twilight glow shown in Figure 7-1 were made by Marjorie starting with the Agung glows in 1963 and extending through 1967. Each point is the average for 10 days. In making such averages one must eliminate poor glows caused by cloudiness at the tangent point. A good indicator of such cloudiness is the presence or absence of reddened sunlight on the glowing arch. Note the minimum brightness in spring and the paucity of strong glows. The glows recover in intensity in summer, but with an interesting difference. The fall and winter glows tend to show a distinct upper edge to the glowing arch, indicating a rather sharp upper boundary of the aerosol layer. The summer glows have a soft upper

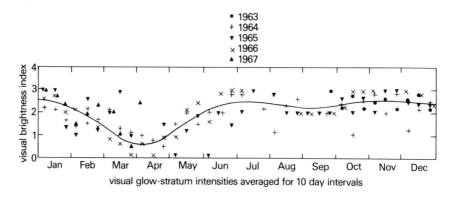

Figure 7-1. Visual estimates of the brilliance of the sunset glows over several years, showing a seasonal drop during spring and enhancement the next summer and fall.

edge. Further, the summer displays often show a strong purple light, but the golden arch develops very little. We suspect that the summer displays are "ordinary" sunsets as discussed in Chapter 4 and not ones caused by an enhanced aerosol concentration at 15 to 20 km (50,000 to 66,000 ft). This suspicion is supported by two facts: first, the purple light appears and disappears earlier than do the volcanic purple glows, indicating an origin lower than 15 km; second, summer convection in the desert is vigorous, carrying dust and haze much higher than in winter. It is therefore probable that the summer glow intensities noted in Figure 7-1 are not caused solely by the sulfate aerosol layer. Because the September glows did show the re-formation of the aerosol layer, we think there is a seasonal change in the layer's ozone concentration.

Another aspect of the glow stratum is its spectrum. After Krakatoa there was some debate about whether the glow resulted from emitted light or scattered sunlight. Observations with a visual spectrometer quickly settled the question. We repeated this type of observation using an interference wedge, a simple form of spectrometer. Held to our eyes, this strip of interference colors enabled us to see that the rain bands in the spectrum were quite strongly absorbed, indicating the presence of much water vapor along the line of sight (Plate 7-1). We detected four bands, at wavelengths of about 6,530, 6,320, 5,900, and 5,760 A (Angstrom units), from red to yellow-green. These strong bands are compatible with sunlight passing low over the Pacific Ocean before illuminating the aerosol stratum over California, whence we in Arizona observed the scattered sunlight close to our western horizon.

ORIGIN OF THE AEROSOLS

What exactly causes the glows? We know it is sunlight scattered off
something resembling dust in the high atmosphere. If the "dust" is
ash, why did the glows return? For the correct answer, a number of
pieces of the puzzle had to fit together. Our observations made us
wonder if the glows really were caused by the ash itself. Once ash
settles out, what can reinject it, short of a new eruption? Remember,
too, that Krakatoa and Agung produced a layer at 20 km, and we had
noted that Surtsey also appeared to have added ash to the 20-km level.
How could volcanoes of vastly different degrees of violence inject matter to the same altitude? We published our hypothesis that the glow
stratum was not ash, except possibly briefly after the eruption, but was
an aerosol of chemical origin – smog.[2] (A friend had told us that the
U-2 "meteorological" planes at nearby Davis-Monthan Air Force Base
in Tucson had, during high-altitude flight, acquired a sticky deposit
on their windshields that burned the hands of groundcrewmen who
cleaned the planes' exteriors.) We noted that some years earlier C. E.
Junge had shown that a sulfate aerosol layer exists at all times at an
altitude of about 20 km. Could the volcanic aerosol layer at 20 km be
nothing more than an augmented Junge layer? We thought so.

If this hypothesis was correct, a volcano had merely to inject into
the atmosphere a large quantity of sulfur dioxide, which would slowly
diffuse upward until it reached the base of the ozone layer at 20 km
where a chemical reaction with ozone would produce sulfuric acid.
The acid would condense on the tiny, solid condensation nuclei that
are always abundant in the atmosphere. As the drops grew, they would
slowly rain out of the sky until the sulfur dioxide content was depleted
to its equilibrium level.

We were delighted to receive a letter on this subject from S. C.
Mossop in Australia:

I was interested to read your joint article . . . in *Science* of 13 January . . .
There seems to be strong evidence that at an early stage some of the volcanic SO_2 is transformed to H_2SO_4. Our stratospheric samples (U-2) con-
tained blobs of water-soluble fluid as well as volcanic dust. This fluid caused
marked corrosion of the copper microscope grids. The windshields of the
returning U-2s were coated with a viscous fluid. We scraped some of this
off and had it analyzed and found that it was almost entirely H_2SO_4.

Mossop further noted: "By about a year after the Agung eruption all
the particles larger than 1 micron diameter had disappeared." Yet the

65

glows returned – because the re-formation of ozone in the 20-km region occurred and new sulfuric acid droplets formed. In view of this mechanism every volcanic eruption does its bit to cause enhanced sunset glows. The amount of enhancement depends on the contribution of sulfur dioxide, not necessarily of ash.

Perhaps the Agung glows were special because the material had traveled so far to reach our northern latitudes. El Fuego and, most recently, El Chichón gave us further evidence on this question. A visible cloudlike structure to the "ash" stratum made us expect that a major component found by U-2 sampling flights would be volcanic ash. But the samples disclosed mainly liquid droplets of sulfuric acid on tiny condensation nuclei. These nuclei included silicates that could have been of volcanic origin, and in the case of El Chichón they also included salt crystals. Nevertheless, the prevailing view now is that outside the immediate region of the eruption the dominant component of the visible veil is droplets of aerosols, mainly sulfuric acid.

If the Junge layer and the volcanic eruption clouds we observed prior to El Chichón were the same manifestation of an aerosol layer at the base of the ozonosphere, what about the El Chichón clouds seen at altitudes of 25 to 30 km (82,000 to 98,000 ft)? The answer seems to be that the eruption injected material as high as 30 km, and this material was soon converted to the aerosol by the richness of ozone above 20 km. If so, this initial injection should soon rain out and be replaced by a more stable layer at 20 km until all the excess sulfur dioxide from this eruption is removed from the atmosphere. We shall see.

VISUAL MEASUREMENT OF STRATUM HEIGHTS

Even more information can be obtained using only the eye as the measuring instrument. The length of time after sunset that the glow stratum can be seen can be easily converted into altitude of the stratum. The geometry is simple. In Figure 7-2 we have added to the sunset diagram quantities we know and quantities we wish to calculate. The equations are also simple when expressed as reasonable approximations to the exact equations. The basic quantities are shown in Figure 7-2. The observer sees the last reddened rays of the sun shining on the stratum at an angle, Θ, above the horizon at a time, Δt, in minutes, after sunset. The angle, α, of the sun below the horizon

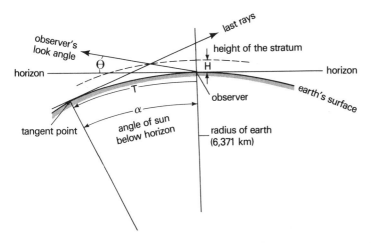

Figure 7-2. Geometry used in deriving the equations for determining the height, H, *of the stratum and the distance,* S, *to the tangent point.*

depends on the latitude, λ, of the observer and the time after sunset, according to the approximate expression

$$\alpha = 0.25 \, \Delta t \, \cos \lambda.$$

For high latitudes where the summer sun dips only a modest distance below the northern horizon this equation is not good, but for latitudes of about 50° the approximation is reasonable all year (see Chapter 2).

If you can estimate when the upper edge of the glow reaches the horizon, the equation for the height, *H*, is given by

$$H = \frac{R}{2} \tan^2 \left(\frac{\alpha}{2} \right),$$

where R is the radius of the earth (6,371 km or 20,903,520 ft). When the local sky is hazy, estimating the time of glowset is difficult, although with experience it can be done to the nearest minute. By noting that the upper edge is dropping toward the horizon at a uniform rate and extrapolating this motion, one can arrive at a reasonable glowset time. The graph in Figure 7-3 allows you to convert your observations directly into the stratum height without calculations. The dash-line example shows how a time from sunset to glowset of 37 minutes yields a height of 15 km (50,000 ft) at our latitude of 32° (dashed-line curve). You must remember that this height is that above where the sun's rays are tangent to the earth, so if there are clouds at

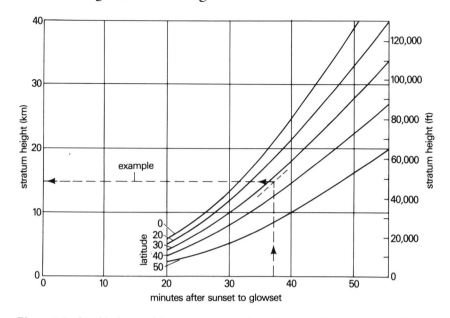

Figure 7-3. Graphical way of determining the height of a volcanic stratum from the time after sunset to glowset on the horizon, at various latitudes. The example is for a time of 37 minutes after sunset as seen at the latitude of Tucson (32°).

the tangent point you must add the altitude of these clouds to get the stratum height. The same thing applies if a mountain range is at the tangent point or if you are already at a high altitude, for example in Colorado. You must add the appropriate altitude to get the height above mean sea level of the stratum. Fortunately for us in Tucson, most of the year the tangent point is over the Pacific Ocean.

Because it is difficult to determine the glowset time on the horizon, we have prepared the graph in Figure 7-4 for observations when the glow reaches an angle of 10° above the horizon; here the dividing line between the pink of the glow and the blue twilight sky is easier to see. It is not easy, however, to estimate angles above the horizon without some sort of aid. A convenient gnomon is your hand. Hold out your hand at arm's length and turn your palm toward you. The span of your hand at the palm, as illustrated in Figure 7-4, is close to 10°. If you want to calibrate your hand, simply measure the distance, D, from your eye to your hand and divide the span, S, of your palm by this distance. The angle is given by

$$\tan \Theta = \frac{S}{D}.$$

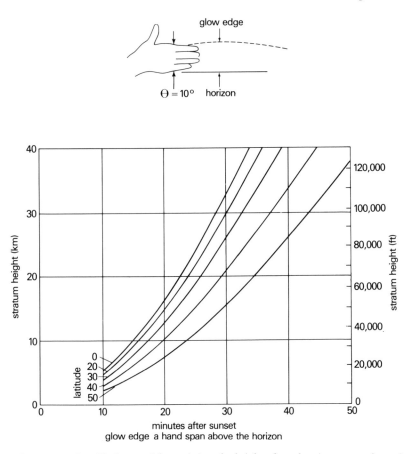

Figure 7-4. Graphical way of determining the height of a volcanic stratum from the time after sunset to when the glow edge is 10° above the horizon. The hand is a convenient gnomon to estimate the 10° angle.

The graph in Figure 7-4 translates into height the time after sunset when the upper edge of the glow reaches the top of your hand. You still need to add the height above sea level at the tangent point.

If you want to know where the tangent point is, the equation for calculating this distance, T, is given by

$$T = R \tan \alpha = 0.25\, R\, \Delta t \cos \lambda.$$

The relationship between minutes after sunset to glowset and distance to the tangent point is given in Figure 7-5. The dashed lines give an example typical for Tucson, where a time of 44 minutes is found, by following the dashed lines, to be 1,040 km (640 mi).

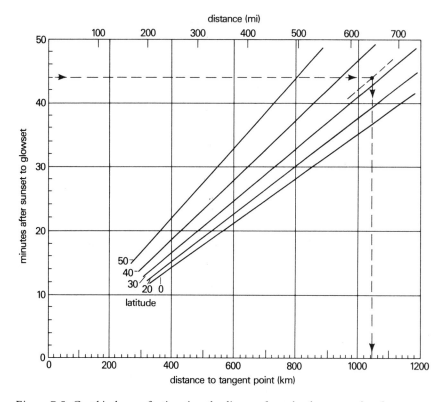

Figure 7-5. Graphical way of estimating the distance from the observer to the solar tangent point as a function of the time after sunset to glowset. The example for Tucson (32° latitude) shows that for a time of 44 minutes the tangent point is about 1,040 km (640 mi) away, over the Pacific Ocean.

PHOTOMETRIC OBSERVATIONS

The observations previously discussed are mainly descriptive, with the science only of a quantitative nature. Determining altitudes on the basis of time of glowset involves only the eye, a clock, and a little geometry. The method is precise enough to show the great height of the glow stratum. Determination of time of setting of the glow stratum is only an estimate for two reasons: first, atmospheric absorption renders the glow invisible when it lies on the horizon; and second, the upper edge is not sufficiently precise, often being so diffuse that it blends gradually into the blue of the late-twilight sky above the glowing arch. Our visual estimates of glowset are always an extrapolation based on the rate at which the glow appears to be setting while still clearly visible above the horizon. Our time estimates are uncertain to the extent of ± 1 minute. It is further almost impossible to gain any

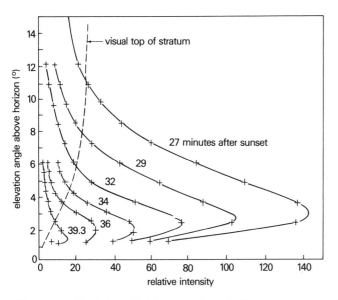

Figure 7-6. Variation of brightness of the twilight glow as measured with the photometer, along with our estimates of the visual top of the glow.

knowledge of how many layers are present or of the distribution of dust. On only one occasion could we derive a distribution; that was from the appearance of a cloud shadow embedded in the glow.

When the twilight glows had persisted so long after Agung, we decided we needed more than visual and photographic observations. We borrowed a photoelectric photometer from Steward Observatory and set up an improvised scanner on our west veranda, as shown in Plate 7-2. Scanning proved to be a two-person task, one setting the vertical angle on the instrument quickly, the other recording angles and visual notes on the chart recorder tracings. Because the glow moved significantly during the 10 angles used, we had to repeat in the opposite direction at measured pace so that the average of two readings at each angle would give a mean brilliance, as though the readings were all taken at a given moment. Some typical records are shown in Figures 7-6 to 7-8.

On several occasions we noted the appearance of a discrete shadow of a cloud surrounded by the glowing stratum, as in Plate 6-2, photographed from Kitt Peak. On one of these occasions, we had the photometer set up; the trace is shown in Figure 7-7. Assuming the distant cloud to be thin and illuminated edge-on by the sun, we can translate this curve as the shadow ray passing upward through the scattering dust and aerosol. By comparing the observed trace with

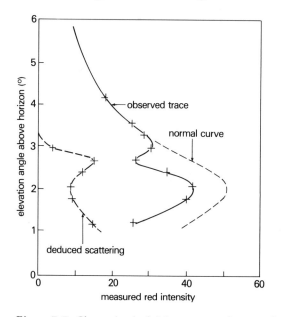

Figure 7-7. Change in the brightness curve due to a cloud shadow. Compare the deduced height profile of scattering with the measured curves in Figures 7-9 and 7-10.

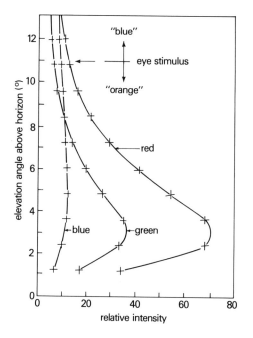

Figure 7-8. Variation of appearance of the twilight glow in different colors. Note there is almost no horizon brightening in blue light.

what occurs in a clear twilight, we can deduce the vertical distribution of the scattering layer, as shown by the dashed curve.

THE EYE'S SENSITIVITY TO COLOR

When you trace the vertical glow intensity, you soon learn something more about the eye. Figure 7-8 shows the change of intensity of red light with elevation angle and time as the glow descends in the west. You now see that there really is no upper edge to the glow. But a distinct upper edge was visible, as noted by the dashed line. This anomaly can be explained when we compare the vertical distribution for three colors: red, green, and blue. The "top" of the glow stratum is where the eye stimulus changes from signaling "orange" and "blue." The photometer's sensitivity to colors is different from the eye's, the eye being more sensitive to green than to either red or blue. Further, the sensitivity of the eye to red changes relative to blue as intensity dims (the Purkinje effect). Finally, the eye loses all color sensitivity at the level of night illumination and sees only shades of grayness as the rods in the eye take over the visual functions from its color-sensitive cones.

The eye's lack of color sensitivity at low levels of light can be dramatically illustrated by appropriate photographs. Stars show conspicuous colors in photographs, as do the star trails displayed in Plate 11-7. There you see the blue stars in the constellation Orion with the violet-pink of hydrogen emission from the Orion nebula near the center. You also can see some very red stars. Looking at this constellation with the eye discloses only the weakest coloration. Betelgeuse, at the upper left shoulder of Orion, does look yellow to the eye with respect to Rigel, at the lower right foot of Orion, but stars of lesser brightness produce little color sensation.

Another scene where the eye does not see much color is a landscape illuminated by the full moon. The camera, on the other hand, sees little difference except in exposure time.

OBSERVATIONS ABOVE THE ATMOSPHERE

When the observer is above the atmosphere, the vertical distribution of haze in the high atmosphere becomes easy to see and measure because the layers can be viewed edge-on against the blackness of space. For this reason, the altitude of volcanic aerosol layers has been measured from photos made by cameras in balloons and in manned satel-

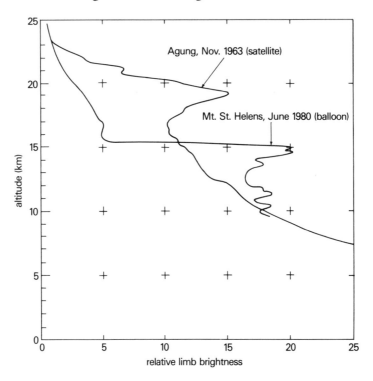

Figure 7-9. Horizon brightness records for volcanic ash-aerosol layers as seen above the limb of the earth. (Balloon data from M. Ackerman, C. Lippins, and M. Lechavallier. Nature *292, 587, 1981)*

lites (Plates 13-1 to 13-4). But these occasions are few and the cost high in relation to the number of observations. From satellite photos one can get essentially an instant view of the haze distribution around the earth between the latitudes traversed by the satellite. Figure 7-9 shows our measurements, made using an early photograph from space, of the volcanic aerosol layers above the limb of the earth after the Agung eruption. The figure shows also a similar type of measurement based on a photo taken from a balloon after the Mount Saint Helens ash had reached Europe. This type of measurement gives excellent height resolution. One would like an equally accurate method that could be used from the surface of the earth.

LIDAR MEASUREMENTS

This better way to measure the vertical distribution of haze above the surface is called lidar, an acronym analogous to radar, denoting a system that uses light rather than radio wavelengths to measure distance.

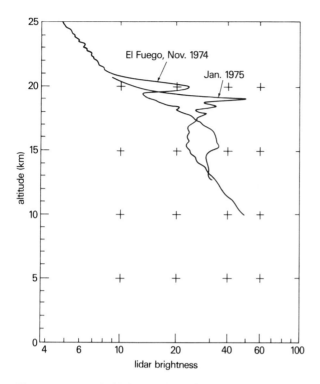

Figure 7-10. Vertical lidar soundings through the atmosphere after the eruption of El Fuego in 1973. (M. P. McCormick and W. H. Fuller, Jr., NASA Langley Research Center)

The development of lasers provided a strong source of light that can be sent aloft as a burst of photons only a few centimeters in length. A 1.2-m (48-in.) lidar system used by M. P. McCormick and his colleagues at NASA Langley is shown in Plate 7-3. It consists of two basic parts: the laser, shown mounted on the upper right side of the telescopes, and the telescope to gather the weak echo of the laser beam returning to earth. A very short pulse of intense laser light is sent aloft. As it passes through the atmosphere, any scattering material in the atmosphere scatters some of this light back down. The telescope is equipped with a photomultiplier to detect this faint echo, and the signal output as a function of time after the upward burst of light is triggered yields the distance to the scattering material. The intensity of the echo yields the amount of scattering material, after corrections are made for the diminution of intensity by distance. A typical laser ranging profile of the vertical distribution of volcanic ash from the eruption of El Fuego is shown in Figure 7-10.

Such direct probing shows much detail in the vertical structure of the atmosphere after a major eruption. Because the number of lidar installations in different latitudes is growing, we can learn about the effect of volcanic dust and aerosols and at the same time learn more about atmospheric circulation. When a sensitive probe is available, even the smaller eruptions can serve as atmospheric tracers. Future lidar observations will surely be of growing value to the study of the earth's atmosphere.

LESSER GLOW ENHANCEMENTS

The identification of the origin of a new occurrence of volcanic ash or aerosol is relatively easy when a major eruption such as Agung, Surtsey, or Fuego occurs. It is much more difficult when there is an enhancement but no new major eruption. In a compilation of all eruptions causing damage and casualties (presumably only the large ones), you will find that in some years there are many medium-sized eruptions but no large ones. In other years you will find several large explosions and still a number of medium-sized ones. Taken together these lesser eruptions could collectively contribute enough sulfur dioxide into the atmosphere to make a noticeable effect. Figure 7-11 shows the measured sulfate concentration in the upper atmosphere over a number of years. The Agung eruption was a major contributor. Surtsey was a minor one by comparison, but because it lasted several months it produced a noticeable augmentation of the Agung aerosols. Several lesser peaks are also shown with the main contributors noted above each peak.

We researched all the major eruptions that had made conspicuous changes in the skies back to about 1700 (Figure 7-12). There seems to have been a real lull in volcanic activity from the 1912 eruption of Katmai to the 1963 eruption of Agung. We wonder if an active period like the one in the nineteenth century is now in store for the earth.

EXTINCTION OF SUNLIGHT

The overhead passage of volcanic ash and aerosols is evidenced by photometric measurements of the diminution of sunlight and starlight. Precision measurements give information on the rate at which the ash and aerosols rain out of the atmosphere. In the case of Krakatoa precision measurements were not possible. Even so, observation of sunset glows showed that the sulfuric acid droplets continued to produce effects at least 6 years after the eruption. In the case of Agung we observed progressively weaker glows from 1963 to 1965, when

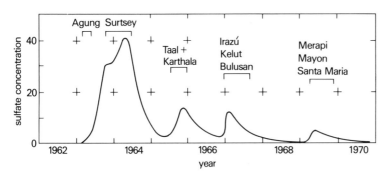

Figure 7-11. Variation of sulfate concentration in the upper atmosphere from 1962 to 1970, and known volcanic eruptions during the same period. (Data from A. N. Castleman, Jr., H. R. Mukelwitz, and B. Manowitz. Tellus 26, 222, 1974)

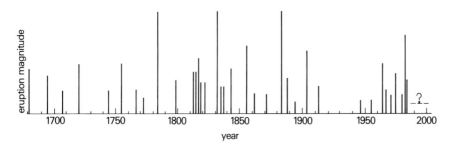

Figure 7-12. Major volcanic eruptions causing reported instances of dry fogs, twilight glows, and other atmospheric manifestations, from 1700 to 1982. Note the quiet period in the years immediately after 1912, until about 1963.

the eruptions of Taal and Irazú added new sulfur dioxide. Measurement of the sulfate particles aloft is a good indicator of the reduction with time, as shown in Figure 7-11. Note that a tail extends at least 2 years after the Agung eruption, until Taal injected new material.

The 1912 eruption of Katmai strongly affected the northern hemisphere and hence was well observed. F. E. Volz compiled the quantitative observations of the changes in atmospheric turbidity (extinction) after this eruption. The findings, displayed in Figure 7-13, indicate an exponential decay with time, with effects lasting 2 years. During this period highly colored twilight glows were reported in the United States and Europe. We have noted that the twilight glows continue after the extinction observed at Kitt Peak has returned to "normal," which indicates that the aerosols still scatter enough to be seen at twilight but have very small effect on bulk absorption.

Many authors have speculated on the possibility that large volcanic eruptions can make a noticeable change in the climate in the following

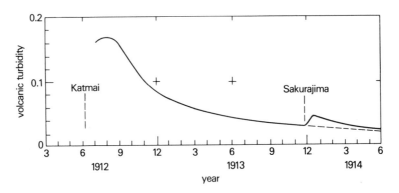

Figure 7-13. Variation in atmospheric turbidity after the eruption of Katmai in 1912. (Data from F. E. Volz. Journal of Geophysical Research 80, *2643, 1975)*

years. Measurements of extinction at astronomical observatories show there is little reduction in the passage of starlight, except for a few weeks after the eruption – hardly enough to affect the amount of sunlight reaching the surface. The effect on the heat balance of the earth is more subtle, depending on the re-radiation of the heat load, which in turn depends on the types of molecules in the atmosphere and their distribution with height. Geologic evidence indicates that in periods of intense mountain building when volcanism was at a peak the temperature of the atmosphere was changed; thus, there is much scientific interest in trying to detect any small changes that contemporary eruptions may cause.

In this chapter we touch only upon those scientific aspects of twilight that we have personally explored, only a tiny bit of the breadth of investigations now being pursued after the El Chichón eruption. Satellite observations provide new vantage points; so do the extensive probes using high-flying aircraft. El Chichón has produced a big perturbation of the upper atmosphere at a time when the necessary tools are at hand to explore the question, Do volcanic eruptions influence the climate of the earth? We await the findings with much interest.

8

Bishop's ring and blue suns

Besides the already described sunset and twilight effects produced by volcanic eruptions, two other phenomena associated with solid matter in the atmosphere are appropriate topics for this book. They are Bishop's ring and blue and green suns. The solid matter responsible for their occurrence is sometimes of volcanic origin, but may also be desert dust or industrial pollutants. For example, the windborne loess dust carried aloft over China from the Mongolian and Takli Makan deserts is one source of such displays; another is the air pollution found in and near industrial centers (far more common before the environmental cleanups of today).

BISHOP'S RING

A strange disk of luminous light fringed with color was seen around the sun in hazy skies over Hawaii on 5 September 1883 by the Reverend S. E. Bishop. He wrote:

Permit me to call special attention to the very peculiar corona or halo extending from 20° to 30° from the sun, which has been visible every day with us, and all day, of whitish haze with pinkish tint, shading off into lilac or purple against the blue. I have seen no notice of this corona observed elsewhere [to that date]. It is hardly a conspicuous object.[1]

Our drawing of Bishop's description is presented in Figure 8-1. The halo was quickly termed Bishop's ring and soon many observers around the world reported sighting a similar ring, but with curious differences. These independent sightings confirmed the existence of the ring, but descriptions of its size and appearance varied. Dr. C. Meldrum, F.R.S., who was one of its earliest observers in the tropics, spoke of it as "a whitish silvery patch surrounded by a brownish fringe, with a

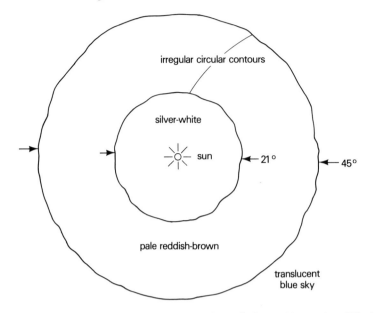

Figure 8-1. Drawing of Bishop's ring as observed after 1883 eruption of Krakatoa, based on Bishop's description. Note that the ring was not precisely circular as a lunar ice halo would be; moreover, the diameters of the inner and outer boundaries varied with time, location, and observer, probably as the size of the scattering particles varied.

radius of 12° to 24°, according to the position of the sun."[2] Other viewers described the color inside the reddish-brown ring as bluish. The greatest disagreement, however, concerned size. Some observers reported a diameter of only 14°, others 80°, but most gave about 45°, the inner edge of color being 21°. The ring was not necessarily seen as a true circle, often being reported as irregular in outline (Figure 8-1).

It must be remembered that in September 1883 most of the world, especially the tropical belt, was covered with the ash clouds from the Krakatoa eruption. Such clouds are distinctly different from weather clouds and are a challenge to description. Photographs cannot convey the feeling one gets looking for the first time at volcanic ash clouds. Look at the three examples in Plates 6-16, 6-22, and 6-25, and then look at some of the descriptive words used in the Krakatoa report: very high filmy cirrus, disposed in transverse bands or ripples, very delicate in form; wavy rippled silvery haze; striated, fine clouds, but not really like cirrus; soft delicate cirrus; recurved waves and bars. Bishop described the ash cloud as a misty rippled surface of haze, always perfectly transparent, invisible except under certain conditions.

It is apparent that Bishop's ring was not seen when daytime ash clouds were still present, but only later, when the residual material was an almost invisible veil. We have continually looked for a similar ring phenomenon, but have seen it only twice. The heaviest sky obscuration from an actual ash cloud passing over us occurred after the Volcan de Fuego event (Plate 6-15). We had an opportunity to see what looked very much like a Bishop's ring while the El Fuego ash was aloft. In addition to the ash, there were some thin clouds in the sky that contributed to the formation of a silvery disk with a reddish tint at the boundary. We photographed the ring using a saguaro cactus as a sun shield and got the picture in Plate 8-1. To the eye the glare of the silvery inner area was so bright that we did not notice the greenish coloration inside the reddish ring until we examined the photographs.

The ash from El Chichón in 1982 also was associated with a faint Bishop's ring. We saw the halo, and several others reported it to us and to the Smithsonian. It was most often described as a brownish ring around the silvery central region.

There are occasions when a similar phenomenon can be seen in thin clouds around the sun. In these instances the diameter of the ring is much less, indicating that the scattering particles are larger than in the case of a volcanic cloud. Uniformity of size is also more readily achieved with water droplets formed around condensation nuclei. When the ring occurs in clouds, it is usually referred to as the solar aureole. Bishop's ring is technically only an aureole, but of much different angular diameter and associated with a cloudless sky – a sky with only a thin veil of volcanic dust and aerosol.

A water-cloud aureole showing the silvery disk and orangish fringe is shown in Plate 8-2. The cloud was a high cirrus layer; a jet flying through it a few minutes earlier had cleared a path, but the path was much distorted at the time the photo was made.

ICE HALOS AND RAINBOWS

It was noted by several physicists in 1884 that Bishop's ring is not a *refraction* ring, as is an ice halo, but a true diffraction ring. Ice crystals form a perfect ring halo around the sun or full moon – a halo that is 22° in diameter and with red innermost, blue outermost. For comparison Plate 8-3 shows a photograph of an ice halo around the full moon. The best seen ice halos are the ones around the moon; those around the sun are consistently paler. We have often wondered about this situation and think that the heat of the sun leads either to more turbulence during the daytime or to a lesser equilibrium abundance of the hexagonal-prism ice crystals necessary for refracting light into a 22° ring.

Whereas light refracted through ice crystals produces the 22° ice halo, light reflected and refracted through droplets of water produces a rainbow. Rainbows are often associated with sunset inasmuch as the sky is then clearing after an afternoon summer shower and the bow is arched high in the sky (Plate 8-4). For a complete pictorial exposition on the many types of rainbows the reader is referred to *Rainbows, Halos, and Glories.*[3]

On rare occasions the rainbow phenomenon is seen at night, caused by the light from the full moon. This bow is a ghostly one. It has been described as the black rainbow. The reason for this description is that at low brightness the eye loses its sensitivity to color. You see gradations of color brightness as shades of gray.

SMOKE CLOUDS VERSUS WATER CLOUDS

Water and ice clouds are basically different from ash clouds in two important respects. First, weather clouds are continually giving up or absorbing heat energy as a consequence of the heat of vaporization or freezing. This heat exchange causes vertical motions that help shape the characteristic forms of clouds. Volcanic ash clouds are composed mostly of inert, nontransparent solids. Only where there is saturated water vapor does water form on these solid particles to modify their gross appearance; this is probably what produced the Bishop's ring shown in Plate 8-1.

Second, volcanic ash or aerosol clouds are desiccated; the material floats serenely on the relatively stable airflows high in the atmosphere. The soft, rippled silvery haze structure seen in volcanic clouds thus looks quite different from the more defined boundaries of weather clouds. Condensation of sulfuric acid on ash nuclei may also play a part in shaping the volcanic ash clouds' appearance.

The solid ash particles only scatter sunlight because they are opaque. Whereas refraction effects produce rainbows and halos of fixed diameter, diffraction effects depend on the size of the particles involved. Small particles scatter at larger angles from the sun than large particles. When particles of all sizes are present in about equal numbers, the sum of their diffraction effects is a general haze whose intensity drops off at a uniform rate with angle from the sun. To produce a Bishop's ring effect, most particles must be of the same size. In this case one will see the color effects resulting from diffraction by this particular size. Because large particles drop out of the atmosphere faster than small particles, the size of ash particles remaining aloft reduces with time. One would then expect Bishop's rings to vary in size with distance from time after the volcanic eruption. The ring associated with Fuego was smaller in diameter than the ring seen after Krakatoa's eruption because larger particles were involved after El Fuego than after Krakatoa.

BLUE AND GREEN SUNS

Blue and green suns are other diffraction phenomena. They were widely seen after Krakatoa in regions where the ash clouds were so heavy they almost obscured the sun. Whereas the green flash is an instant of vivid coloration occurring the moment the smallest segment of the sun is visible on the horizon, the green and blue suns are entirely colored, generally including a very pale whitish tint, as long as the clouds causing them continue to obscure the sun.

Shortly after the Krakatoa eruption, observers (mainly in the tropics) saw the sun appear blue, green, silvery, yellowish, and coppery and the moon occasionally green. Even before the main eruption, blue suns were noted. The log of the ship *Elisabeth* reported: "On the morning of 21st [May 1883] the [quality of the] light was that which prevails during an eclipse of the sun [i.e., faint]; the sky presented the aspect of a large dome of very thin opal glass, to the vault of which the sun seemed suspended as a pale blue globe."[4] After the main eruption on 27 August, at Batavia (Jakarta) the sun was observed to be green after emerging from the cloud of smoke of the eruption.

In September the sun in India was seen at midday to have a greenish-blue tint, dimmed by haze, and turning bluish toward 4:00 P.M., with banks of smoky haze across the solar disk as it set. That same month in Ceylon (Sri Lanka) green suns were noted, the coloration extending around the sun into the ash clouds. In February 1884 in San Salvador the crescent moon was seen 15° above the horizon at sunset as a magnificent emerald green on an immense crimson curtain; also, a small comet in the sky was clearly green, as was Venus.

This last observation of strong color is an important one for the following reason. The response of the brain to visual color sensations is highly subjective. Edwin Land has spectacularly demonstrated that in the absence of normal color cues the eye can "see" a full range of colors when in fact the scene is composed of only yellow to red wavelengths. The brain apparently is sensitive to slight color differences, and when normal colors are absent these small wavelength differences are expanded into a broader spectrum. In the case of the San Salvador observation, the moon, the comet, and Venus shining on a vivid red background could easily appear green even though they were quite white.

There are, however, true physical effects that produce colors in transmitted light, and these are generally thought to account for appearances of green and blue suns when the sky background is neutral gray clouds composed of small particles. Such unusual colorations of the sun and nearby sky can be noted when dust of a particular size is suspended in the air. Blue or green suns and moons have been observed at times when the atmosphere is dusty or hazy. The cause can be, for example, industrial smoke (often reported in industrial England in the days before environmental cleanups) or a loess dust storm over China.

Dr. Peter Franken, director of the Optical Sciences Center where we work, had an opportunity to photograph a blue sun when he was visiting Beijing, China, in 1978. As he came out of a building after a meeting, he noticed a reddish tint to the landscape. His Chinese hosts casually remarked that it was only one of the dust storms that occur when the fine loess dust is swept aloft over China from the deserts of Mongolia. Peter naturally looked up to the sun and remarked with some surprise that the sun looked bluish. Again his hosts explained that the sun frequently was blue when this dust arrived. The Royal Society's report on Krakatoa remarks that Baron von Richthofen "in his work on China notices that the air in Central Asia is filled with dust, and that the sun seen through it appears merely as a dull bluish disc."[5] Fortunately, Peter had his camera and took the pictures shown in Plates 8-5 and 8-6. Plate 8-5 shows a pale bluish tint to the sun and the nearby clouds. Peter says the sun was actually even more colored than in the photograph. This is because the color recorded on film is sensitive to exactly the correct exposure, and the sun was overexposed making the colors paler. Plate 8-6, taken pointing the camera 90° away from the sun, shows the reddish tint to the scene and sky.

Why was the sun blue but the scene orange? One would normally expect a reddish scene to be illuminated by a reddened sun, but here was a blue sun lighting the landscape red! This situation is rare, except in places such as China where windborne dust travels a long distance before arriving overhead, but it is readily explainable by optical physics. It is a manifestation of the Mie effect first discovered by G. Mie in

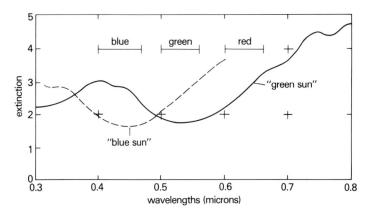

Figure 8-2. Variation in particle absorption as a function of color according to the Mie scattering theory. The solid curve is for a situation producing a greenish sun; the dashed curve is for smaller particles, producing a blue sun.

1908. He showed mathematically that a small spherical particle scatters light preferentially with wavelength and direction, the extinction varying sinusoidally with light frequency when the wavelength of light is about the same size as the diameter of the particle.

A typical variation of extinction (scattering) with wavelength for a dust cloud of single-sized spherical particles is shown in Figure 8-2 for Mie scattering. The exact position of the green minimum depends on the index of refraction of the dust grains (usually between 1.5 and 1.6), on the size of the grains, and to a much lesser extent on their shape. Particles 1.1 μ diameter (0.00004 in.) produce a green sun; smaller ones, 0.85 μ (0.00003 in.), produce a blue sun. Even for a cloud of single-sized particles the coloration is not strong inasmuch as the scattering varies only by a factor of 2 over the visible spectrum. Thus when a distribution of many sizes is present, the colors are subtle. Note that for the green or blue sun of Plate 8-5, the red scattering is high. This explains why the sun and the sky close to the sun are bluish (Plate 8-5), whereas the same scene (Plate 8-6) photographed at right angles to the sun is reddish.

We, too, saw the dust-veiled sky over China during our stay in Nanjing in 1979. It was in November, before the heavy dust veils arrive, but even so there were days when we could comfortably look at the sun well before sunset. When we remarked how easy it was to look at the sun, Dr. Y. C. Chang, emeritus director of Purple Mountain Observatory, smilingly explained that this is why Chinese astronomers discovered the 11-year sunspot cycle over 1,000 years before Galileo observed spots on the sun with his small telescope.

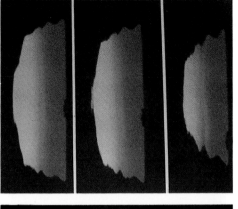

Plate 2-3. The Chinese-lantern effect of inhomogeneous air layers. (O'Connell and Treusch)

Plate 2-2. Oblateness of the setting sun viewed normally and viewed with 90° rotation.

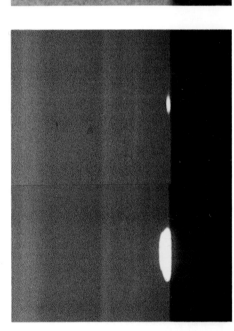

Plate 3-2. a, b: A green-flash sunset sequence (David Cortner); c: a good example (Dennis diCicco). Note the orange sky helps the eye see a greenish tint to the rim of the sun.

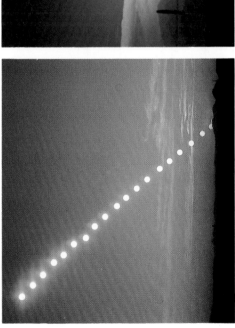

Plate 2-1. Time sequence of the setting sun, showing dimming and reddening.

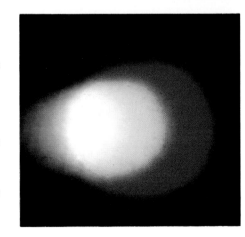

Plate 3-1. Atmospheric refraction as Venus nears horizon, through a refractor. (O'Connell and Treusch)

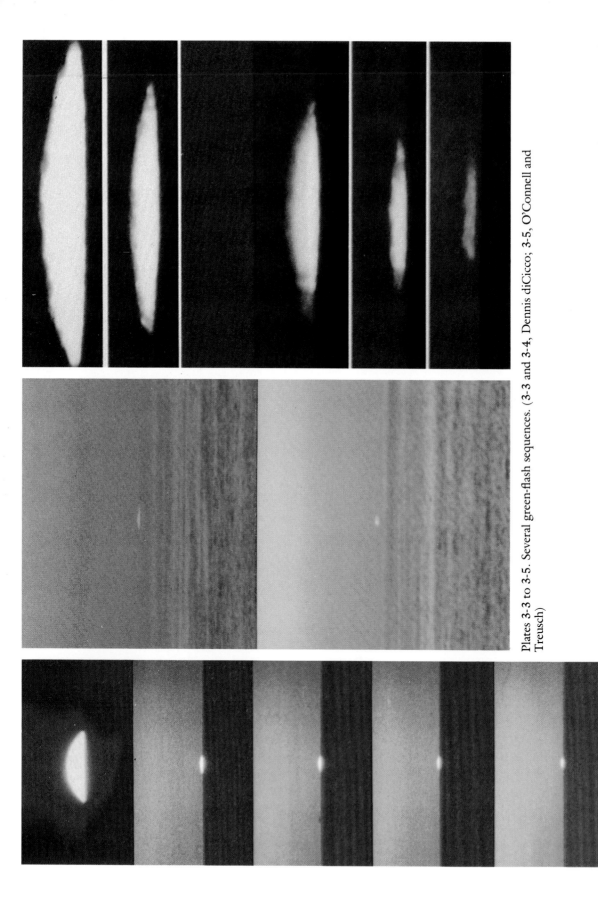

Plates 3-3 to 3-5. Several green-flash sequences. (3-3 and 3-4, Dennis diCicco; 3-5, O'Connell and Treusch)

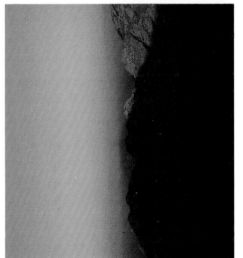

Plate 4-1. Beginning of the earth's shadow below a full moon.

Plate 4-2. Earth's shadow at maximum development. Note pink color at its boundary.

Plate 4-3. Earth's shadow on desert dust over Saudi Arabia.

Plate 4-4. Mountain shadow of Kitt Peak showing the typical pyramidal shape. (W. G. Livingston)

Plate 4-5. Kitt Peak shadow with spike of the 4-m telescope building.

Plate 4-6. Beginning stage of sunset ray development. (Jeannine V. Lamar)

Plate 4-7. Spectacular sunset colorations and rays during a summer desert storm when skies were clear to the west of the storm. Note the contrast between the pink rays and the blue enhanced by the black of the high cloud canopy. (Michael Yada, *Arizona Daily Star*)

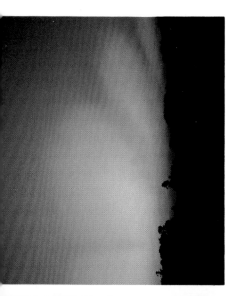

Plate 4-8. Cloud shadows at the time of sunset. Note convergence to the sun.

Plate 4-9. Beginning of crepuscular rays in a clear blue sky.

Plate 4-10. Cloud shadows cast on a high incursion of Mongolian dust over Taiwan, enhanced by pink sunlight.

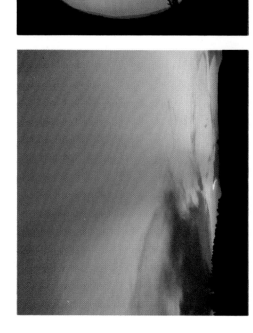

Plate 4-13. Antisolar crepuscular rays seen in Caracas, Venezuela. (Agnes Meller)

Plates 4-11 and 4-12. Crepuscular rays diverging from the west (4-11) and crossing the sky, converging in the east antisolar point (4-12).

Plates 4-14 to 4-16. Development of sunset colorations. 4-14: Lowest clouds in red sunlight. 4-15: Lowest clouds in shadow; high cirrus pink.
4-16: All clouds in shadow; cirrus invisible; a trace of purple light in the twilight sky.

Plate 4-17. Maximum normal coloration of a clear sky at sunset.

Plate 4-18. An occurrence of the purple light, a subtle mixing of scattered red light and sky blue.

Plate 4-19. Late twilight with exceptionally clear sky. Note the rocket-flare trail near the clouds.

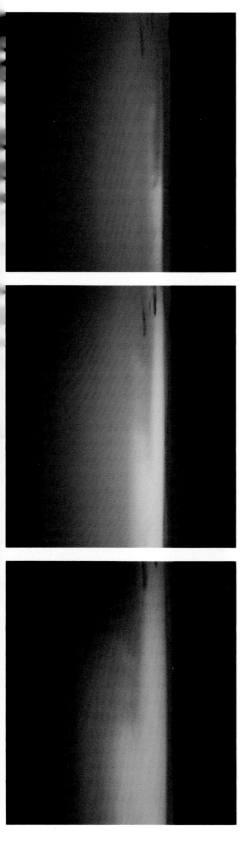

Plates 6-1 to 6-3. Development of the Agung volcanic glow stratum, 20 October 1963, seen from Kitt Peak National Observatory. Note the cloud shadow on the brilliant glow.

Plates 6-4 to 6-6. Agung glow seen under hazy sky conditions from Yerkes Observatory, Wisconsin, 25 October 1963. Note the visibility of the setting of the primary glow to the horizon.

Plates 6-7 to 6-9. An Agung glowset and development of the secondary glow, 19 October 1963. The exposure time on each frame is considerably increased. Compare the secondary coloration with Bishop's description.

Plate 6-10. Clouds silhouetted on the purple light of an Agung glow, 11 October 1963.

Plate 6-11. Strong Agung glow seen even against Tucson city lights, 6 November 1963.

Plate 6-12. Agung glow seen from Boston, 25 November 1963.

Plates 6-13 to 6-15. First sighting of the El Fuego ash stratum and its subsequent development, indicating its high altitude, 21 November 1974.

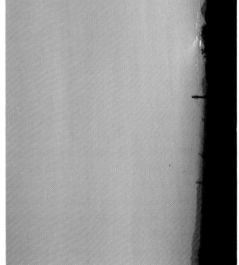

Plates 6-16 to 6-18. Daylight-visible ash stratum at sunset, 22 November 1974, and its development, showing ripple structures and a coloration sequence similar to the Agung glows.

Plates 6-19 and 6-20. Two Christmas twilights: 6-19, in 1968 after the Agung glows had subsided; 6-20, in 1974 during the El Fuego glows.

Plate 6-21. Glow with red sunset illumination cut off by a cloud layer over the horizon.

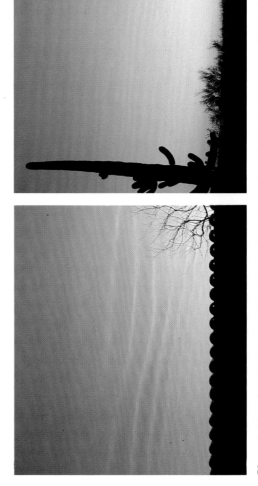

Plates 6-22 and 6-23. Passage of the Mount Saint Augustine ash clouds, 25 January 1976: 6-22, mid-afternoon; 6-23, after sunset.

Plate 6-24. Maximum development of the glow stratum after the Mount Saint Helens eruption, 1980.

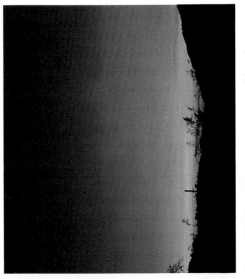

Plate 6-25. Early view of the El Chichón ash cloud, 5 May 1982, showing much structure.

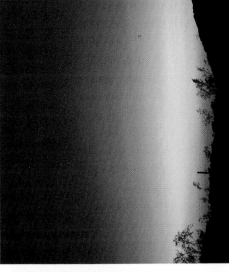

Plate 6-26. Blockage of the glow above sunlit ash clouds, 6 May 1982.

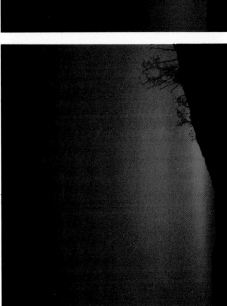

Plate 6-27. Partial transparency to reddened sunlight, 15 May 1982.

Plate 6-28. Homogeneous but double shadow bands indicate double ash layer; ugly color compared with Agung.

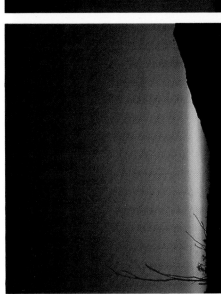

Plates 6-29 and 6-30. Setting of the shadow band caused by self-absorption of the ash stratum for solar rays tangent to the layer, 2 July 1982.

Plates 6-31 and 6-32. Maximum development of the El Chichón glows seen from La Jolla, California, 21 July 1982. The higher altitude of this volcanic stratum is shown by the intensity of the secondary glow (6-32) 60 minutes after sunset.

Plate 6-33. Twilight glow stratum seen from Mount Wilson Observatory, overlooking the Los Angeles area, with crescent moon. (James K. Heck)

Plate 7-3. NASA Langley lidar telescope and control vans. (NASA–LRC)

Plate 7-2. Improvised scanning photometer for measuring Agung glows.

Plate 7-1. Looking at the Agung glow through an interference wedge.

Plate 8-3. A 22° ice halo about the full moon: a refraction halo.

Plate 8-2. Aureole in cirrus clouds with an old vapor trail and a new one with an associated shadow.

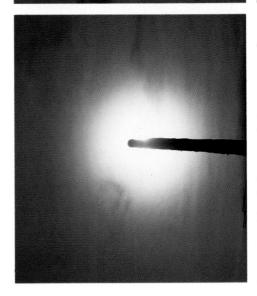

Plate 8-1. Bishop's ring seen during passage of the El Fuego ash stratum. Note colorations.

Plates 8-5 and 8-6. Left: Blue sun seen through a veil of Mongolian desert dust in Beijing, China. Right: Scene 90° from left, showing dust-reddened illumination in Beijing, China. (Peter Franken)

Plate 8-4. Rainbow at sunset: a reflection phenomenon. Tucson.

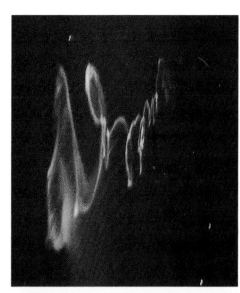

Plate 9-3. Convoluted rocket trail seen from El Centro, California. (David Drach-Meinel)

Plate 9-2. Rocket trail seen in late twilight from Tucson.

Plate 9-1. A high-latitude noctilucent cloud over College, Alaska. (Yngvar Gotaas)

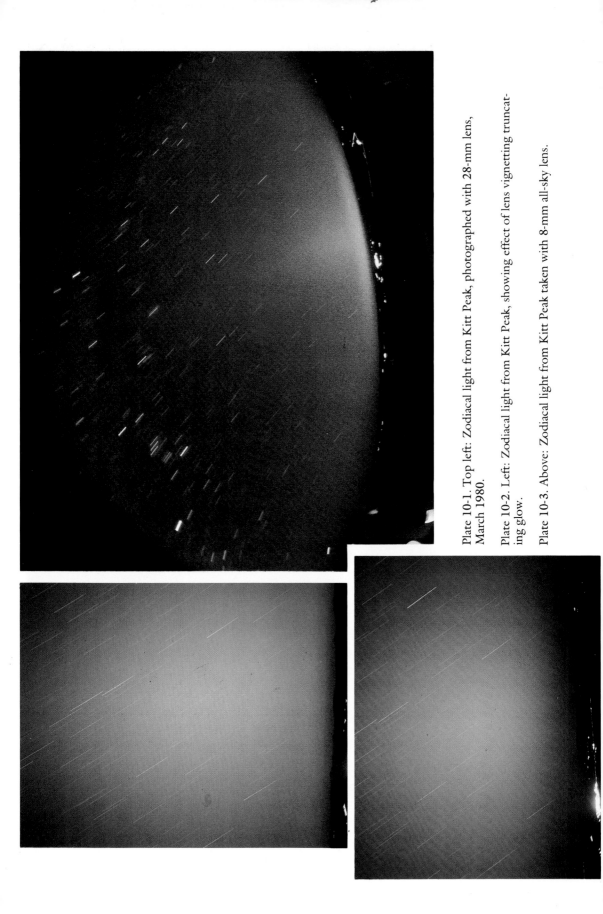

Plate 10-1. Top left: Zodiacal light from Kitt Peak, photographed with 28-mm lens, March 1980.

Plate 10-2. Left: Zodiacal light from Kitt Peak, showing effect of lens vignetting truncating glow.

Plate 10-3. Above: Zodiacal light from Kitt Peak taken with 8-mm all-sky lens.

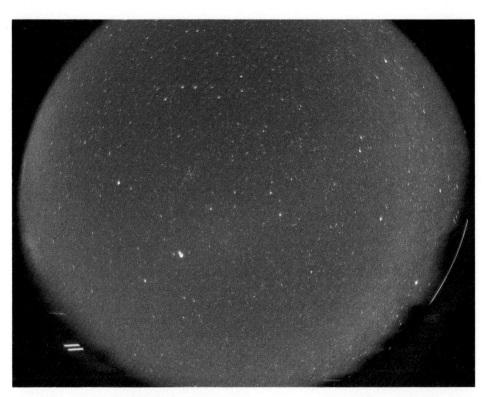

Plate 10-4. All-sky photograph in search of the gegenschein. It lies to the left of Jupiter and Saturn, but is of very low visibility.

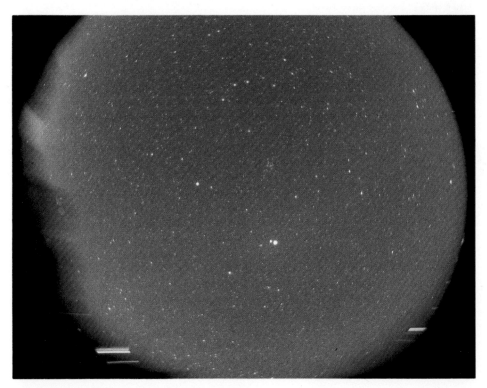

Plate 10-5. All-sky photograph in search of the gegenschein. It is now centered on Jupiter and Saturn.

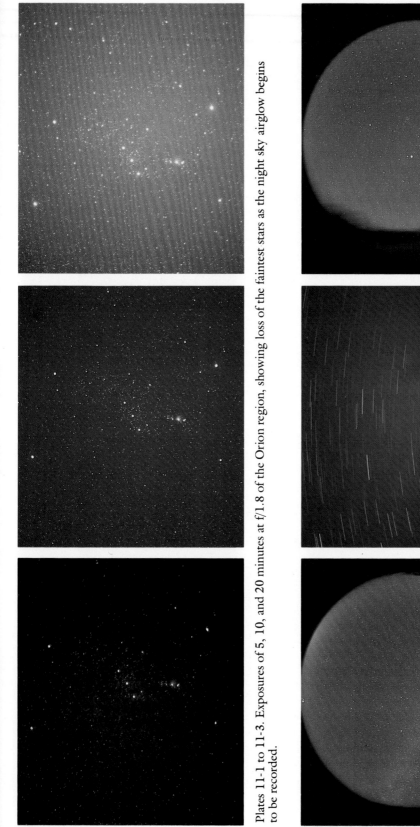

Plates 11-1 to 11-3. Exposures of 5, 10, and 20 minutes at f/1.8 of the Orion region, showing loss of the faintest stars as the night sky airglow begins to be recorded.

Plate 11-6. All-sky photograph showing airglow brightening at horizon.

Plate 11-5. Airglow brightening near south horizon from Kitt Peak.

Plate 11-4. The Milky Way and Orion with zodiacal glow (right) and Tucson glow (left).

Plate 11-7. Orion star trails showing the coloration of many stars.

Plate 11-10. Bluish auroral rays converging toward the magnetic zenith in a display seen overhead. (David A. Huestis)

Plate 11-9. An intense auroral storm. (David Cortner)

Plate 11-8. Pink auroral rays on a green glowing background. (David A. Huestis)

Plate 11-13. A low-latitude red auroral glow. (NASA–JPL Table Mountain Observatory, James Young)

Plate 11-12. Blue-rayed aurora. (Philip L. Dombrowski)

Plate 11-11. Red-rayed aurora. (Dennis Milon)

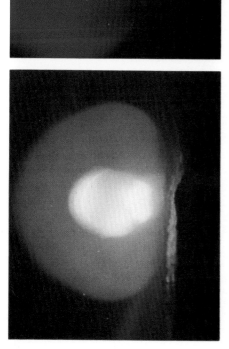

Plates 11-14 and 11-15. Development of intense airglow emission after the atomic explosion Teak, as seen from Hawaii. (John Champeny)

Plate 11-16. The tropical aurora caused by the detonation of Starfish, as seen from Hawaii. (Herman Hoerlin)

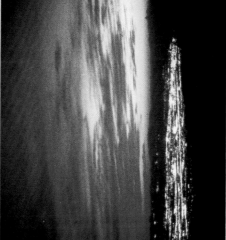

Plate 12-1. City glow of Tucson seen from Kitt Peak National Observatory, 1981.

Plate 12-2. Tucson at twilight from the Catalina Mountains. (David Drach-Meinel)

Plate 12-3. Twilight with city-illuminated clouds, Tucson.

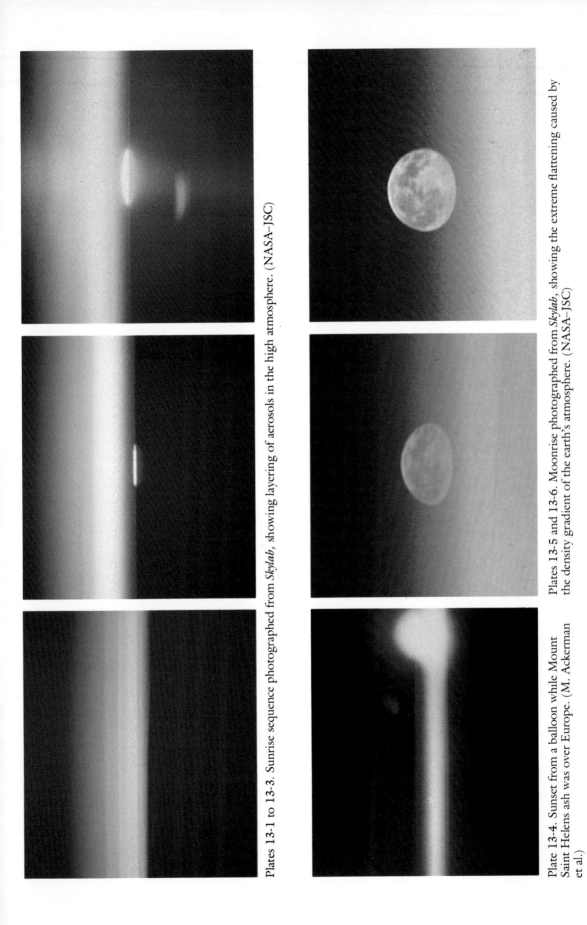

Plates 13-1 to 13-3. Sunrise sequence photographed from *Skylab*, showing layering of aerosols in the high atmosphere. (NASA–JSC)

Plates 13-5 and 13-6. Moonrise photographed from *Skylab*, showing the extreme flattening caused by the density gradient of the earth's atmosphere. (NASA–JSC)

Plate 13-4. Sunset from a balloon while Mount Saint Helens ash was over Europe. (M. Ackerman et al.)

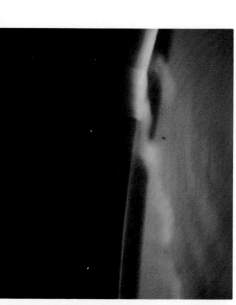

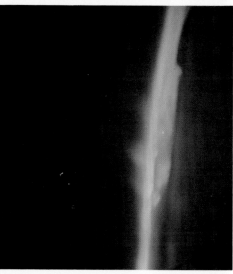

Plate 13-7. Airglow layers seen from *Skylab*, greenish on top, reddish below, with twilight bottom. (NASA–JSC)

Plate 13-10. Sunset through fine dust on Mars as recorded by *Viking*. (NASA–JPL)

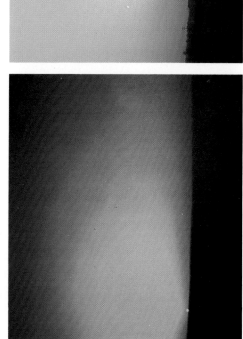

Plates 13-8 and 13-9. Two views of the aurora australis, 11 September 1973, photographed from *Skylab*. Note the reddish layer extending left of the aurora in 13-9. (NASA–JSC)

Plate 13-11. Setting sun through heavy dust, Nanjing, China, showing self-shadowing near the ground.

Plate 13-12. Jovian satellite Io with eruptions seen at "twilight." (NASA–JPL)

Plate 14-3. A man-made comet: a vapor trail at sunset, with crescent moon above. (David Drach-Meinel)

Plate 14-2. Comet West (1976) seen in New Hampshire. (Betty and Dennis Milon)

Plate 14-1. Comet Ikeya-Seki (1965) rising over an Agung sunrise glow. (Richard H. Cromwell)

9

Noctilucent clouds

NATURAL CLOUDS

Noctilucent clouds are clouds that remain sunlit long after sunset. This indicates that they are at a very great height. Natural noctilucent clouds are rare, occurring only at high latitudes in the late summer, and are therefore seldom reported except from northern Europe, Canada, and Alaska. They do not look like the soft-featured volcanic clouds, but closely resemble weather clouds. Plate 9-1 shows a spectacular specimen, photographed from the Geophysical Institute at College, Alaska. There is some twilight color near the horizon and some dark clouds in the foreground, but the noctilucent cloud is still fully illuminated by uncolored sunlight and thus must be high in the atmosphere.

These clouds remain visible long enough after sunset and are observable from locations sufficiently far apart that their altitude can be estimated by simple surveying triangulation. The height is found always to be close to 80 km (50 mi); thus they are far above weather clouds and volcanic dust.

The origin of these natural clouds has been debated. Because they are at about the height where most small meteors burn up, they were once thought to be meteoric dust, but in appearance they resemble water-vapor clouds. Their most probable origin becomes apparent when one looks at the physical situation of the atmosphere at 80 km. A very low temperature minimum is found at this altitude (Figure 9-1). The temperature drops to the order of $200°$ K ($-100°$ F). We have observed a temperature of $170°$ K ($-135°$ F) for this altitude in a study of an infrared oxygen-emission band in the night sky. This temperature is so low that the small amount of water vapor that occasionally penetrates above the stratosphere to this height freezes out into an ice cloud.

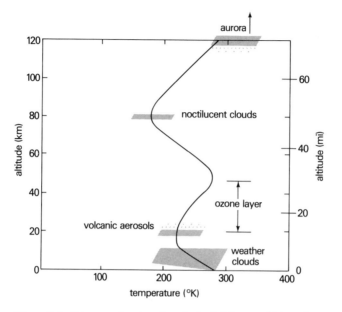

Figure 9-1. Diagram showing the relative heights at which various atmospheric phenomena occur and the temperatures at these heights.

MAN-MADE CLOUDS

In view of the rareness of natural noctilucent clouds, we were surprised to begin seeing very high clouds in the northwestern skies of Tucson. We had finished photographing the Agung glow stratum on 2 November 1963, and had the tripod and camera in hand, when our daughter Carolyn noticed a noctilucent cloud appear right where we had been looking only a few minutes before. We photographed it. Then twice again, on 13 and 20 December 1963, noctilucent clouds appeared. One of these clouds is shown in Plate 9-2. Note that the twilight sky is quite dark, only a trace remaining of the colors, but the cloud is still in unreddened sunlight. It does not show the same type of structure as the high-latitude noctilucent cloud in Plate 9-1.

This series of photographs enabled us to measure motion, which, with altitude derived from time of sunset on the cloud, yielded ground velocity. We projected backward and found the time and place the cloud must have been when it was over the Pacific coast. It was exactly the time when a rocket had been launched toward the Kwajalein Island impact area from the place we had determined: the Vandenburg Missile Range in California. Since then we have occasionally seen other clouds, sometimes even the luminous rocket exhaust expanding like a gossamer balloon against the dark sky, followed by a lasting trail.

Recently, our son David, driving across the desert near El Centro, California, after sunset, saw in the western sky the dramatic spectacle of a rocket-exhaust noctilucent cloud. Fortunately, he had his camera in his car. Plate 9-3 shows the sunlit trail, much convoluted by the winds at various altitudes traversed by the rocket.

Artificial noctilucent clouds have also been produced by the ejection from rockets of fluorescent or chemically luminous material at high altitude, usually to make high-altitude winds visible. These clouds are frequently colored red or green, depending on whether barium or lithium chemicals are used. An example is shown in Plate 4-19. A high-altitude research projectile was fired from a special gun at the army's proving grounds at Yuma, and a chemical tracer that fluoresced in sunlight was ejected into the sunlight zone, where it remained for about 20 minutes, gradually diffusing and drifting, as shown in the lower part of the picture.

METEOR TRAILS

A meteor, when large enough, may leave a visible trail of smoke or luminous material behind, but this disappears in a few minutes. Microscopic meteors also continually rain into the atmosphere. They provide condensation nuclei for water and ice droplets, form clouds, and so could become instrumental in also forming ice clouds in the thermal minimum at 80-km altitude.

10
Zodiacal light and the gegenschein

ZODIACAL LIGHT

One of earth's most serene sights is the western sky in late winter after the last colors of twilight have left the western horizon. The stars are out, and if you are far from the lights and haze of cities they are brilliantly etched on a black sky. So many stars are visible that even the bright stars marking the constellations are almost lost among them; but there, marking the ecliptic, is the zodiacal light – soft and white, blending imperceptibly into the night sky itself. Some of our best views of the zodiacal light have been in west Texas – Big Bend country. The observatory houses at McDonald Observatory in the Davis Mountains are in that perfect country for observing the zodiacal light. The only man-made light one sees is an occasional faint gleam from the adjacent observatory buildings or a lone car on the road from Marfa to Fort Davis. Above the mountains to the west rises the wedge of luminosity, absolutely serene.

From most locations the zodiacal light is difficult to see. You need to know when to view it best and what it looks like. As the name indicates, it is a band of light aligned along the constellations of the zodiac. In other words, it lies along the ecliptic – the path the sun follows throughout the year. The plane of the ecliptic is inclined 23°.5 with respect to the plane of the earth's equator. This is why the sun is high in the sky in summer, low in winter. When the planets are in the sky they are strung like beads along the ecliptic, sometimes a bit above or below, because the orbital planes of all the planets are close to the ecliptic. The ecliptic also coincides very nearly with the plane of rotation of the sun.

Photographing the zodiacal light is not easy if your aim is to show it as you see it. Its vertical extent is most conspicuous in winter because the Milky Way does not interfere as it does later in the spring.

As a consequence, the zodiacal light holds center stage in the west. At about 75 minutes after sunset (at Kitt Peak) it becomes apparent above the last traces of twilight. Some 20 minutes later it is at its best, a luminous tapering band of light, bright at the horizon, then gradually fading as it reaches nearly to the zenith.

In looking at pictures of the zodiacal light, we were disappointed to find they did not much resemble its visual appearance. We decided to see if we could get better pictures than others have obtained, using Kitt Peak as the location. We knew that an ordinary 50-mm lens would have too small a field of view to cover much of the soft wedge of the zodiacal light, so we used a 28-mm wide-angle lens. The result is shown in Plate 10-1. The last trace of the sunset glow colors the western horizon, above which we see a band of atmospheric absorption. Above that band is the delicate soft color of the zodiacal light, but it hardly looks like a wedge stretching up beyond view at the top of the picture. The wedge appears to be extinguished by a general darkening of the sky. Look closely and you will also see that the stars seem to have disappeared near the edges of the field of view. This darkening is a universal problem of lenses (called vignetting), especially when a lens is used at its fastest f-stop setting. A second exposure with the same lens is shown in Plate 10-2, the way most people take a picture; here the vignetting even more severely truncates the top of the zodiacal light.

We remembered that the Flandrau Planetarium on campus had a special wide-angle lens made for the purpose of taking movies for projection on the dome of the planetarium: an f/2.8 lens of 8-mm focal length. Such a lens should produce no vignetting over a wide enough field to allow us to photograph the zodiacal light as it really looks. The result is shown in Plate 10-3. This record is much closer to what the eye sees – a white wedge of faint light stretching far up into the sky. The exposure was 15 minutes and the camera fixed, so the stars appear to move. The zodiacal light also moves as the sun sets, so it too is blurred, but the light is so diffuse that the blurring does not affect its appearance. If we placed the camera on a star-tracking mount, the stars would appear as points of light and the horizon blurred. We tried several exposures of this type, but the blurred horizon prevented the picture from showing the zodiacal light as you would see it.

Even though we overcame the problem of vignetting, the zodiacal light shown in Plate 10-3 still does not appear quite as it does to the eye. This is because of the logarithmic response of the eye, discussed in Chapter 2. The eye can see the band as a slowly tapering wedge of light reaching relatively high in the sky, almost to the meridian at the onset of astronomical night. Film has a linear response to brightness which, when combined with the small dynamic range of tone possible in a color print, makes the zodiacal light appear differently.

The photographed color of the zodiacal light depends on the film used, the lens, and the exposure time, as shown in Plates 10-1 to 10-3. Although the spectrum of the zodiacal light is close to that of yellowish sunlight, the visual color is white. This is because the eye cannot register color of extremely faint light.

The zodiacal light was once thought to be sunlight shining on the very high atmosphere of the earth; it has now been shown to be sunlight reflected off debris left over from the formation of the planets. Most of this debris lies close to the plane of the sun's rotation and rapidly trails off in density on either side of the ecliptic. The color of the zodiacal light is essentially the same as that of sunlight. This means that the debris is not gaseous atoms or molecules, because these are so small the Rayleigh scattering would produce a sky-blue glow. Particles much larger than the wavelength of light, therefore, must produce the bulk of the zodiacal light, and we assume that they can range from micron-sized dust grains to meter-sized miniasteroids. The rain of small meteors is thought to be made up of typical objects that together cause the zodiacal light.

The tilt of the earth's orbit causes the apparent path of the ecliptic to be tilted with regard to the projection of the earth's equator on the sky, varying with the time of year. Figure 10-1 (top) shows a view of the western horizon for 21 March with the path of the ecliptic and general outline of the zodiacal light. At the spring equinox, the light is most nearly vertical and hence best seen. At the fall equinox it lies close to the horizon and the light is lost in the atmospheric haze (Figure 10-1, bottom). For best predawn views of the zodiacal light the situation is reversed: the fall equinox is best.

Because the zodiacal light is an interplanetary, not an atmospheric, phenomenon it does not set as rapidly as does the twilight glow. Photographing the latter demands fast action to achieve exactly the desired sunlight configuration. The zodiacal light, however, sets slowly in the west at the same diurnal pace as the stars. You can wait until the last traces of the sunset glow have disappeared at the western horizon before beginning your exposure to photograph the zodiacal light. The last visual evidence does not disappear in late winter until about 2 hours before midnight.

The wedge shape of the zodiacal light results from a nonuniform

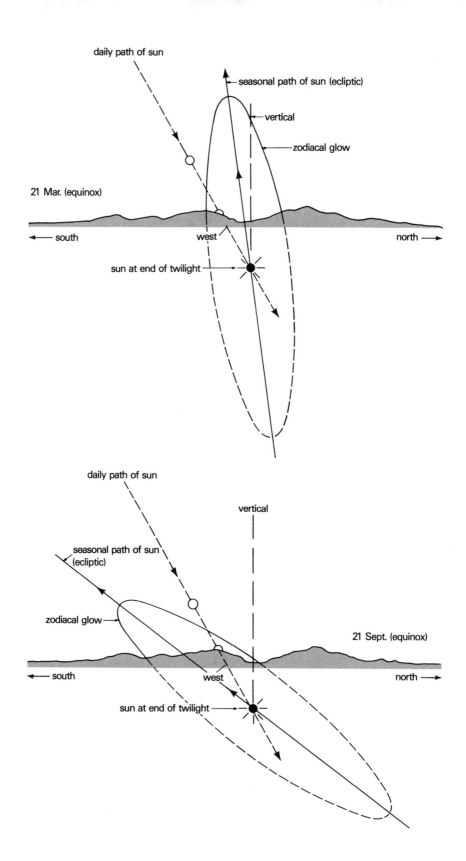

daily path of sun

seasonal path of sun (ecliptic)

vertical

zodiacal glow

21 Mar. (equinox)

← south

west

north →

sun at end of twilight

daily path of sun

vertical

seasonal path of sun (ecliptic)

zodiacal glow

21 Sept. (equinox)

← south

west

north →

sun at end of twilight

distribution of interplanetary material. It is densest at the sun, but extends outward to beyond the orbit of Mars, diminishing rapidly with distance. An astronaut near the orbit of Mercury would see a bright band of light reaching across the sky. One near Jupiter would see nothing except a luminous coronal haze close to the sun, appearing as a glowing ring within which the sun is embedded. In fact, because it is a continuous distribution of dust, the zodiacal light as viewed from the earth merges into the corona of the sun.

The zodiacal light actually stretches across the whole sky – faint, but detectable by the polarization of its reflected and scattered sunlight. The earth is embedded in this band of interplanetary material, a band made more evident by meteors. If one takes the observed rate and luminosity of meteors and calculates their size before they entered the atmosphere, their reflectance, and the volume of space swept out each day by the earth, the result just about accounts for the brightness of the zodiacal light.

HUNTING THE GEGENSCHEIN

The fact that some interplanetary material lies beyond the orbit of the earth means that we should see a brightening in the antisolar direction. This gegenschein, or counterglow, is an extremely faint brightening, a diffuse patch in the sky. It should be, because each miniasteroid or meteoroid is illuminated face-on, in the mode of the full moon. This means that every crater, pit, or crack is unshadowed, and the maximum return of light results. The same principle causes the heiligeschein (holy glow) that you can see from an airplane when flying low over some terrains, such as fields of wheat stubble. Around the shadow of the plane you see a brightening because all the stubble stalks are illuminated face-on. Just a small angle off this antisolar direction part of the stalks and the adjacent ground are shadowed, and the combined luminosity quickly is lowered. The gegenschein is the most difficult to see of all the visual sky phenomena.

The reason for the elusiveness of the gegenschein is that it must be seen against a sky background of maximum blackness. If it occurs anywhere near the Milky Way, it will be lost in the mottled light of these distant star clouds.

Figure 10-1. Position of the zodiacal light (top) at the vernal equinox, when it rises nearly vertically in the western sky after twilight has ended, and (bottom) at the autumnal equinox, when it is poorly visible.

We had doubted that the gegenschein could be seen except by someone who knew it should exist. In science highly motivated seekers often think they have found something, only to be disappointed when others are unable to confirm the find. This was our position when we began our search for photographic proof of the gegenschein. When we examined photographic evidence, such as Figure 3-17 in the NASA publication *Skylab's Astronomy and Space Sciences,* we saw that the number of stars decreased with distance from the center of the picture just as did the glow that was thought to be the gegenschein. This suggested that the glow might be no more than the effect of vignetting of the astronaut's lens.

We, and Marjorie's father, had often looked for the gegenschein from McDonald Observatory, where the blackest skies are found. We had not seen it. Aden had looked for it with Fritz Zwicky in the High Sierra of California (as described in Chapter 11), but they could not see it.

Inquiring of some of our colleagues here at Tucson, we got some encouragement. Nicholas Mayall, famed for his visual acuity, said he could not be certain, but thought he had seen the gegenschein by averted vision from Cerro Tololo Inter-American Observatory in Chile. Harlan Smith reported seeing it from McDonald Observatory. Arthur Hoag said he had probably seen it some years ago from the Naval Observatory in Arizona. Working at the 1.0-m (40-in.) telescope, which has a roll-off roof, he had been curious because a faint cloud kept appearing in the east during the first half of the night. He thought it was just a cloud until he noted that it was at about the antisolar position. However, he has not seen it in recent years because of either the increase of light from Flagstaff or a true variation in brightness of the gegenschein.

Pursuit of the gegenschein led us to set up a small Goto polar tracking mount on Kitt Peak. With the drift of the stars compensated for by this unit, we could freeze the zodiacal glow and stars but have a blurred horizon. About an hour before midnight we took the 1-hour exposure of the zenith sky shown in Plate 10-4. Do you see the gegenschein? A month later we took another exposure, shown in Plate 10-5. Compare the two pictures. In Plate 10-4 the gegenschein is to the right of the bright pair of objects, Jupiter and Saturn. In Plate 10-5 it is centered on these planets. If you stand away from the book you may be better able to see this faint patch of light. If you look very carefully, you may see a faint band of light sloping across the pictures. This is

Figure 10-2. Contrast-enhanced black-and-white photo showing the zodiacal band extending across the sky along the ecliptic. Tucson glow is at the left and the main zodiacal glow at the right.

the extension of the zodiacal light even beyond the earth's orbit. The planets, being also aligned along the ecliptic, lie in this band of light.

Black-and-white film can be easily contrast-enhanced, so we took several exposures on Tri-X film. Boosting the contrast to a very high level yields the picture shown in Figure 10-2. The zodiacal band is now easily seen. The horizon skyglow is very conspicuous; on the left is the city glow from Tucson and on the right the zodiacal glow. The gegenschein was just rising above the Tucson glow at this time.

When we gave a slide show using some of the pictures in this book, we gained a new insight into the visibility of the gegenschein. One of our colleagues asked us to defocus the color slide of Plate 10-4. When we did, the soft spot of the gegenschein immediately became clearly visible! The eye apparently is attracted to the sharp stars in the scene while ignoring the soft background. This may be why placing these two pictures at a distance helps you to see the gegenschein glow – you are less conscious of the stars. This also may explain why a person with perfect vision and acuity may not detect the gegenschein in the sky whereas someone with less perfect vision may be able to see it. Unfortunately, as we write these last words on the topic we cannot experimentally confirm this because the gegenschein in December is superimposed on the winter Milky Way and hence totally lost in the heightened luminosity of the diffuse, glowing starry clouds of the Milky Way.

STAR SCINTILLATION

A glance at a cold, clear winter night sky is sufficient to reward one with the sight of twinkling stars. What causes the stars to twinkle? Why are the dancing and flashing stars even more spectacular in winter than in summer? They twinkle because the atmosphere is a turbulent mixture of air layers of differing temperature. These temperature differences cause small changes in the index of refraction of air, which slightly deviate the rays of starlight as they pass through the atmosphere and irregularly bunch them. When the rays are bunched closer together, the star momentarily appears brighter; when bunched farther apart, fainter. The twinkling is much less noticeable in a telescope because the aperture diameter is much larger than the eye and averages out these effects. The large number of turbulent cells of brighter and fainter intensity, as visible over the aperture of a 2.2-m (90-in.) telescope, is shown in Figure 10-3. To see them, simply remove the eyepiece of the telescope and place your eye directly at the focus of a bright star. The entire aperture of the telescope will appear to be filled with starlight modulated by rapidly fluctuating cells of changing brightness, sometimes barely visible and at other times with as much contrast as in Figure 10-3. These ghostly shadows usually show a systematic motion in one direction, the direction of the wind aloft.

In winter the temperature contrast between air masses is much larger than in summer, so the twinkling effect is different. Also, some bright stars rise in early winter evenings, giving a good opportunity to see twinkling at its best. When Sirius, the brightest star, is rising in the southeast, one can even see flashing colors that, as in the nursery rhyme, make the star look "like a diamond in the sky." Its twinkling colors can be so bright that Sirius has been reported as a signaling unidentified flying object (UFO). Because fainter nearby stars do not show these flashes of color this can seem to be a reasonable conclusion, especially if one is hoping some night to see a UFO.

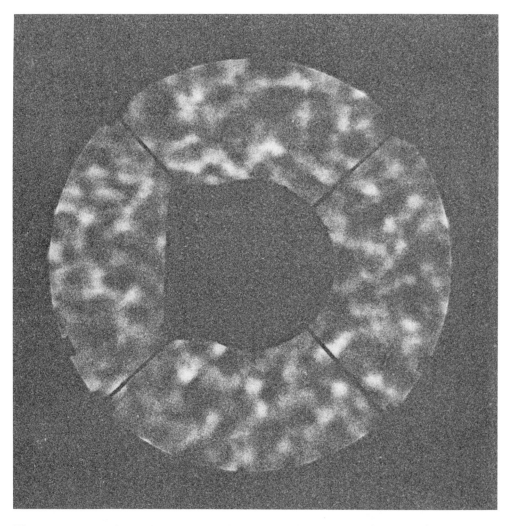

Figure 10-3. A 1/30th-second exposure of shadow blobs caused by thermal turbulence in the atmosphere as seen in the aperture of the Steward Observatory 2.2-m telescope on Kitt Peak on a night of poor seeing.

Steller scintillation is only part of a problem that the astronomer calls "seeing." Scintillation affects the brightness of a star. The other part of seeing is apparent star motion, discernible only with a telescope. The turbulent blobs of air move the stellar image about on the photographic film, blurring the sharpness of the image. In a very big telescope the motion is not observed because of the averaging effect of the aperture; hence a short exposure with a very large telescope looks very much like a longer exposure with a smaller telescope.

If a very short exposure is taken of a bright star the image consists of many tiny bright knots, knots that average out on a long exposure to form a soft star image. These bright knots, termed speckles, are essentially tiny images of the star as seen by the full resolving power of the telescope. By properly processing these speckle images with a computer astronomers have reconstructed the appearance of the surface detail of a few stars, such as the supergiant star Betelgeuse. A whole new field of astronomy, called speckle interferometry, has developed, turning the old problem of atmospheric seeing disturbances into an exciting new research field.

To minimize seeing and scintillation effects, astronomers locate their telescopes on high mountains because most of the atmospheric disturbances are below the observatory site and the air is very clear. From 1955 to 1958 Aden spent much time and effort measuring the quality of the astronomical seeing at a number of sites before recommending Kitt Peak as the best in the southwest for construction of the National Observatory. Many times from Kitt Peak the stars appear steady, but the absence of twinkling does make the sky appear somewhat artificial, as in a planetarium. To an astronomer heading toward a telescope for a night's work, the steady light of the stars promises a night of excellent seeing, a welcome sight.

II

Light of the night sky and the aurora

We have reached the end of the twilight-sky phenomena, and night has descended. We will now look at two variable aspects of the night sky: the airglow and the aurora. They were the subject of Aden's doctoral thesis, ending his student days; they were also the beginning of some exciting adventures at Yerkes Observatory in the early 1950s.

THE AIRGLOW

Even in the darkest location far from cities the background sky between the stars is not completely black. Stars and galaxies too faint to be seen with the naked eye contribute some light, but even when the largest telescope picks them out as discrete objects the sky background still is not completely black. Astronomers wish it were, because that faint background light is enough to fog their long-exposure photographic efforts and to contribute a false spectrum to their spectrograms of faint distant galaxies and quasars.

About 15 percent of the spectral light of the night sky is the reflected solar spectrum from the zodiacal light. The rest is discrete spectral lines and bands, which are emissions from atoms and molecules in the upper atmosphere. These emissions are of photochemical origin, sometimes enhanced when sunlight strikes the atom; for example, the yellow D-lines in the spectrum of sodium.

The background between the stars is not completely dark to your eye mainly because of a most fortunate gift. The dark-adapted eye is especially sensitive to light of $0.55\text{-}\mu$ wavelength, which is the same wavelength emitted by the atomic oxygen line (0.5577μ) in the night airglow. The visual stimulus from this emission is enough to enable you to read very large type, such as newspaper headlines. It undoubtedly helped our ancient ancestors to escape the dangers of the night.

As astronomers, we have walked in total darkness outside observatory domes, carefully, but without trouble. It is very easy to upset this delicate sight condition. A faint red light is the least upsetting, and for this reason astronomers use faint deep-red illumination inside their domes when reading dials or making notes.

The light of the night sky becomes quite evident when you take a time exposure of the sky. To illustrate this phenomenon, we took an ordinary camera with a 50-mm focal length and an f/1.4 lens, mounted it on a small equatorial tracking mounting, and made several exposures of the Orion constellation (Plates 11-1 to 11-3). A 5-minute exposure on ASA 200 Ektachrome film already begins to show the sky background in the center of the field of view (Plate 11-1). One can even see a faint trace of red hydrogen emission to the left of the three belt stars, emissions that are spectacular when a red filter is used that allows only the hydrogen wavelengths to reach the film. The famous Orion nebula is the pink spot in the chain of stars that forms Orion's sword.

A 10-minute exposure shows more stars and more background illumination (Plate 11-2). Note the swarm of faint blue-white stars in the belt region. A 20-minute exposure (Plate 11-3) begins to show the actual color of the atmospheric emissions that cause most of the light of the night sky – a pale apple-green. Note that many of the faint stars visible in the 10-minute exposure are lost, the limit of detection being close to that of the 5-minute exposure. Of special importance, note the dropoff in sky background and stars near the corners of the photographs. This vignetting is inherent in all fast photographic lenses. When the lens is stopped down – to f/5.6, for example – the vignetting is much less, but a long exposure is then required to reach the stars shown in Plate 11-2.

Orion is also shown in Plate 11-4 along with the Milky Way, but here the sky background is bluish. Faint colors on long exposures are sensitive to the film type and processing. In this case, the film had been treated to enhance blue sensitivity in an attempt to show the zodiacal light band across the sky, but with small success. Can you see the zodiacal light? The brightening at ten o'clock is the glow from Tucson.

The star colors do not really show up in the preceding photographs because they are small points of pale color. Star-trail photographs, on the other hand, do show differences in color, as can be seen in Plate 11-7. Betelgeuse, the orange star at the top left of Orion, is bright enough that even with the naked eye you can see the color. The great

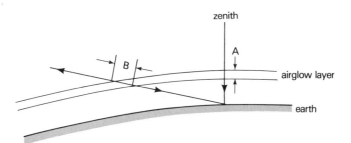

Figure 11-1. Origin of the horizon brightening observed for an airglow layer. When the airglow layer of thickness A is viewed obliquely, the apparent brightness is increased because the path length is increased to B.

nebula of Orion, the pink trail in the middle of Plate 11-7, is too faint: You see it only as a faint whitish star.

Plate 11-5, taken on Kitt Peak (facing due south), shows the brightening of the light of the night sky near the horizon. The sky is reddish, caused in part by the post-twilight enhancement of the red oxygen emission lines and in part by the film or its processing. To avoid the effects of vignetting on the airglow, we took a photograph (Plate 11-6) with the all-sky lens pointed toward the zenith. In this picture the horizon brightening shows as a ring of brightness around the edges of the scene.

Because they arise from layers in the upper atmosphere, the airglow emissions show brightening toward the horizon. The brightening is due to the increased air mass along the line of sight (Figure 11-1). The exact way this light increases has been used by many scientists, especially F. E. Roach and his colleagues, to determine the height of the emissions (the earth's curvature and lower atmospheric extinction are also involved). A typical change of brightness with elevation angle above the horizon is shown in Figure 11-2.

Our direct involvement in airglow studies came about when Pol Swings pointed out to us that the infrared airglow and auroral spectral regions were unexplored. We were at the Naval Ordnance Test Station, China Lake, California – our first desert home, complete with spectacular sunsets over the Sierra range to the west. World War II had just ended, and while Aden awaited the day when he could leave the navy and continue his education, he designed an infrared spectrograph and a new type of camera, which was the flat-field f/0.8 camera of 100-mm focal length. The key element, a transmission grating, was lent by R. W. Wood. Aden built the entire spectrograph and camera system and used it to do his dissertation research at Lick Observatory

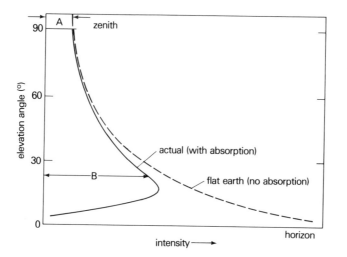

Figure 11-2. Variation of the brightness of the airglow as a function of angle above the horizon. Brightnesses A and B refer to Figure 11-1.

in the following years. It was mounted on the roof of a large water reservoir. All-night exposures were common. The results were unexpected. Infrared photographic materials were very slow in 1948, but even so it became obvious that the infrared airglow was rich in emissions.

One morning when Aden developed the previous night's photograph something quite different was revealed that was to have a major impact on our future work. Some new strong emissions showed close to the northern horizon. Other observers working that night thought they had seen an aurora in the northeastern sky, a very unusual sight in southern latitudes. Aden was able to identify the new auroral emissions as atomic oxygen and to add that finding to a successful doctoral dissertation on the airglow. The airglow emissions still defied identification, except for one pair of lines he identified as being due to the oxygen molecule. Inasmuch as the aspheric plate of the f/0.8 camera had some zonal errors because of the limited time he had in which to polish it, Aden decided to make a new plate of pure fused silicon. Because the aurora had now become a new interest to the government, he received an Office of Naval Research (ONR) grant to continue this work at Yerkes Observatory in Wisconsin where the aurora would not be so rare an event. Further, the top molecular spectroscopist, Gerhard Herzberg, was there.

Arriving at Yerkes we found that Herzberg had left for Canada, but we stayed and set up the spectrograph in the old transit telescope room. The first spectrum recorded with the new camera was fantastic. The lines were clearly etched on the plate, spaced in a regular sequence that Aden recognized was all due to the same molecule. Measuring the spacing, he knew the molecule was OH, a hydroxyl group. How could a minor constituent actually produce the strongest emissions? Aden immediately sent a copy of the spectrum and his thoughts about it to Herzberg. At the same time he started a detailed study of the line strengths, which are the clue to the exact molecular level involved, and decided it was a "doublet pi" level. The unusual thing was that the molecular levels involved were elevated, the strongest being level 9, with 8 fainter, and 7 fainter still. A real puzzle. Aden had also sent his puzzle to Pasadena – to David Bates and Marcel Nicolet, two famous astrophysicists. The answer returned with surprising speed. They had calculated Aden's finding to be a two-body reaction between – of all things – ozone and atomic hydrogen. The energy released was precisely enough to excite the resultant product OH to level 9, not level 10. The other levels were populated by radiative cascades.

News of this discovery spread rapidly. "Here is an organic way to produce oxygen," was Lloyd Berkner's quick response. The reaction was

$$O_3 + H \rightarrow OH(\text{excited}) + O_2 .$$

The original O_3 and hydrogen could arise from dissociation of water vapor by ultraviolet sunlight. Once produced, the natural consequence was a stable molecule of oxygen. This cycle would continuously generate more oxygen and more ozone. Both shield the surface of the earth from excessive ultraviolet light, which kills embryonic life forms (even seawater is transparent to harmful ultraviolet rays). Once the ozone-oxygen shield had been established, the earth was made fit for the appearance of life. Vegetation then could take over the main role in maintaining the molecular oxygen in our atmosphere. By this mechanism our planet, the only one with water and with a temperature such that the water was liquid, was established as an abode for life. Thus discoveries go. Find one new thing and a cascade of discoveries and insights follows.

All this burst of results from Aden's new spectrograph and the identification of OH covered scarcely 3 months. We were still waiting for a good aurora. In the meantime, Aden was offered the vacancy created when Herzberg left Yerkes; he gladly accepted. We hadn't even considered this possibility when we came, hoping instead to return to a California observatory. We moved into an old house – as it happened, a house Marjorie had visited when her astronomer father, Edison Pettit, had worked at Yerkes in the summers of the 1930s. Dr. Pettit had made the first moving pictures of solar prominences, those spectacular clouds of hydrogen gas that erupt from the sun from time to time. We settled down to wait for one of these eruptions to trigger an auroral display.

THE AURORA

One question was foremost in our minds: Do particles, such as hydrogen, reach the earth from the sun during auroral displays? Carl Gartlein had observed hydrogen emissions that were unusually widened. This widening would be the result of hydrogen atoms spiraling around magnetic field lines, but because all Gartlein's observations were to the north one could not be certain. We wanted to view an auroral display by looking directly upward along a magnetic line. Then, if the hydrogen atoms were plunging down into the top of the atmosphere, the Doppler shift would cause the hydrogen emission lines to be shifted to the violet, as diagrammed in Figure 11-3. We thought about going north to Saskatchewan, Canada, where auroras are known to occur overhead, but just 10 months after our arrival at Yerkes, fate stepped in.

One night as soon as it was dark, just at the very end of twilight, we could see that the sky was alive with auroral draperies. They were even overhead. Aden ran all the way from our home to the observatory, while Marjorie hastened to tuck the children in bed before following across the vast observatory grounds. Aden arrived at the transit room only to be confronted with frustration. All the earlier spectrograms had been taken looking north, through an opened north shutter of the transit room. The magnetic zenith where Aden wanted to point the spectrograph was directly overhead. And the long-disused roof shutters wouldn't open. The ropes to open them were over 50 years old, and the roof shutters had been sealed with tar paper so rain wouldn't come in. In desperation Aden pulled hard. Nothing happened. He pulled even harder, and a small ripping sound came from the darkness above. Encouraged, he pulled again and again, praying that the ropes wouldn't break. At last the shutters parted with a loud report, and the blaze of auroral glory filled the open slot. There was a beautiful corona display in the magnetic zenith.

With trembling hands Aden carefully inserted the postage-stamp-sized plate into the camera, always a ticklish task even when he was not excited. He began a 20-minute exposure and held his breath, hop-

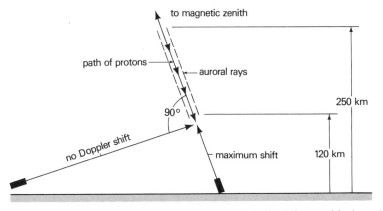

Figure 11-3. Geometry for observing the Doppler velocity shift caused by incoming protons (hydrogen nuclei) during an aurora.

ing the aurora would not fade (something that would often plague him later). This night it remained bright. The exposure over, he pointed the spectrograph to the north to take a picture that would not be affected by any Doppler shift. Then he hurried to the basement darkroom. Fifteen minutes later he burst out and up into the long hallway above, where he saw S. Chandrasekhar. Waving the barely dry plate and a hand lens, Aden excitedly exclaimed to Chandra, "Look, the H-alpha line is greatly shifted to the violet – protons *are* entering the atmosphere at a high velocity!"

Later, careful analysis of those plates confirmed this result and added a puzzle: The velocity was too high to account for the delay between the solar eruption and the aurora. Merle Tuve suggested that perhaps there is some acceleration mechanism near the earth. (This was almost 10 years before Van Allen discovered such a mechanism in the earth's ionization belts.)

Back to that night of the great aurora. After taking several more plates of the proton phenomenon, with the aurora still blazing forth, Aden decided to take one single exposure lasting the rest of the night on the vague hunch that something interesting might show up. We were both emotionally exhausted, and this gave us some time to relax at home. The long exposure did not disappoint us. The usual auroral emissions were uselessly black, but the interesting thing was a curious group of infrared bands that now looked to be clearly related. Aden's work on these puzzles had, however, to wait. The excitement, the chill autumn night, and the exertion of ripping open the roof of the transit room were too much. He ended up with pneumonia. Publication of his discovery of the protons gained national attention. After he had recovered enough, a photographer from *Time* magazine came and took his picture, complete with slippers on.

The analysis of that last exposure produced much new information. Aden found another new band system – this one from a level in ionized molecular nitrogen predicted from theory by R. S. Mulliken, but never found in the laboratory. Here it was in the aurora. With Chung-Yan Fan, Aden continued work on the aurora by reproducing it in the laboratory at Yerkes; with Joseph Chamberlain he studied the airglow and the aurora with a much larger spectrograph, an enlarged copy of his dissertation instrument. He even joined Lloyd Berkner's group planning the International Geophysical Year, in particular the Antarctic auroral program, for the National Science Foundation. The instruments Aden designed went to the Arctic and Antarctic, but Aden didn't. He was suddenly asked to head the search for a national astronomical observatory that 5 years later would result in the dedication of the first large telescopes of the new Kitt Peak National Observatory.

We relate these personal happenings to show how research and exploration in science can be exciting and often unpredictable. Our life has been one unending series of new challenges and opportunities, but throughout it our love of sunset, twilight, and night sky has prevailed. We have enjoyed these scenes in many places on many continents, and some are illustrations for this book.

The usual aurora borealis is a quiet colorless display, most often a soft, glowing arch centered in the north-northeastern sky (the direction of the magnetic pole as seen from the midwestern United States). The magnetic pole itself is located halfway from the top of Hudson's Bay to the north pole, and thus its true direction depends on where you live.[1] This glowing arch may show dark sky beneath and above. The actual color of the arch is greenish, caused by the same oxygen emission as is the light of the night sky, but enhanced about 100 times. This is still too faint to cause a color sensation except for the strongest auroral storms. This quiet arch may remain only a few minutes and then disappear in a glimpse of faint auroral rays. Sometimes the au-

Figure 11-4. (Top) Beginning of the great auroral storm of 19/20 August 1950, taken at the same time that the spectrograph first recorded incoming protons. Note the "corona" effect of the auroral rays converging to the magnetic zenith. Exposure, 10 seconds. (Yerkes Observatory photograph, Stewart Sharpless and Donald Osterbrock)

Figure 11-5 (Bottom) The development of pulsating glowing sheets of the aurora is blended into smooth draperies by the 10-second exposure. (Yerkes Observatory photograph, Stewart Sharpless and Donald Osterbrock)

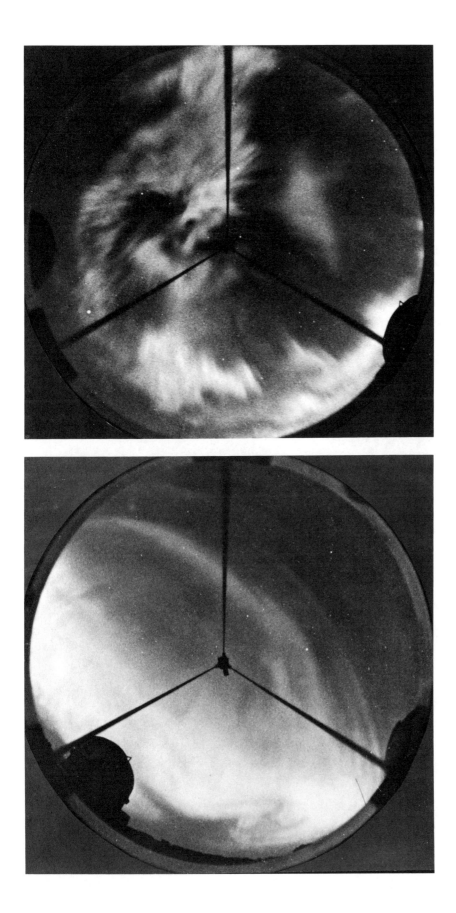

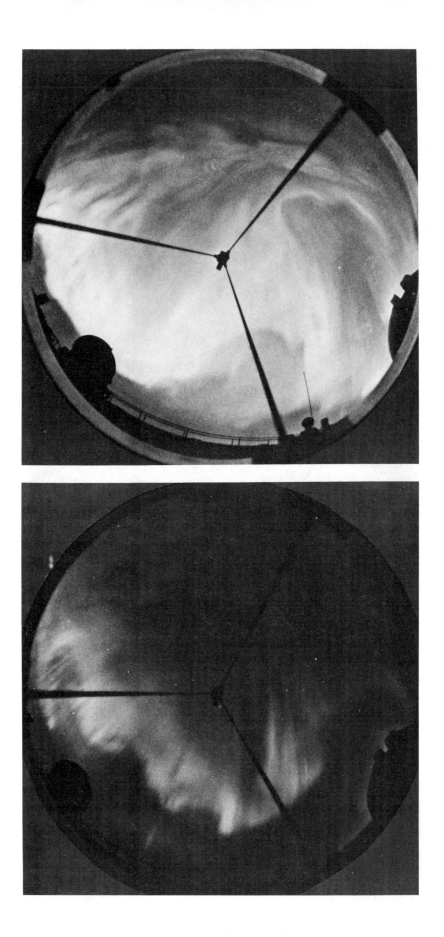

roral display is only a soft, pale red glow, without form, filling the northern sky. Sometimes it is only a sprinkling of transitory razor-sharp rays all pointed neatly along the lines of magnetic force. Rarely do the rays show green color; even rarer is a red portion at their very bottom.

Every few years there is an occasion of a great auroral storm, especially in years near the solar sunspot maximum. These ionospheric storms cause auroras that are bright enough to show color to the eye, but even so, they do not appear as spectacular as they can when recorded with a fast lens on film. In Figures 11-4 to 11-7 we show pictures taken with an all-sky f/2 lens at Yerkes Observatory the same night that we recorded the Doppler-shifted hydrogen emissions. A 10-second exposure time has resulted in integration of fast-changing features into smooth sheets of light. The sky was a blaze of curtains and sheets of innumerable rays, rapidly changing, some fading and others forming every few seconds. The ultimate scene was one in which the entire display pulsated like a theater marquee, alternately brighter and fainter, each pulse showing the same convolutions of structure.

Lenses and films have been improved so that today we can enjoy pictures with color that show more detail. The most common aurora – the quiet arch – is seldom photographed. "Quiet" does not refer to audible effects but to its static visual effect. The pale green arch appears toward the northeast in the United States, the direction of the magnetic pole, lasts for a few minutes or perhaps an hour, and then fades away. At the last stage of a quiet arch a few rays can sometimes be seen, sometimes bright enough to display the colors shown in Plate 11-8. The auroras that gain the attention of camera fans are the more dramatic draperies, usually green, that show changing intensity and shapes (Plates 11-9 and 11-10). The strong red-rayed auroras are uncommon, but striking, as shown in Plate 11-11. Sometimes the rays photograph violet (Plate 11-12), but the difference in ray colors can be influenced by the lens used because of the difference in ultraviolet transmission. Auroral rays are rich in ultraviolet of wavelength 0.39 μ, owing to ionized molecular nitrogen emission. Some lenses transmit some of this light and others none; thus ray colors can vary on the

Figure 11-6. (Top) Maximum development of the aurora, about midnight. (Yerkes Observatory photograph, Stewart Sharpless and Donald Osterbrock)

Figure 11-7. (Bottom) Rayed draperies in the last stage of the aurora, near dawn. (Yerkes Observatory photograph, Stewart Sharpless and Donald Osterbrock)

film. In Plate 11-12 another factor tends to produce a purple tint to the upper reaches of the rays. When the rays reach so high that they are in sunlight, the 0.39-μ emission band fluoresces, adding the extra energy from the sunlight to the emission. To the eye this purple light is invisible, but not to the camera.

Because auroras are seldom announced in advance, many who would like to see a good aurora miss the show. David Huestis, one of our contributors of aurora pictures, remarked on this in a letter to us:

Many amateurs congratulated me on the fine displays I had captured on film. However, they were disappointed because they were unaware an aurora had occurred. Some said I should have called them. But it was late [he worked a second shift and drove home about 11:30 P.M., a good time to see if an auroral show was on]. I explained that I hadn't wanted to upset them . . . but so many of the amateurs indicated a desire to be notified of an aurora that I conceived the idea of the aurora hot line for members of the Amateur Astronomical Society of Rhode Island.

An excellent idea that we are sure many in the northern tier of states would appreciate.

David Cortner, another contributor, wrote us about his experience photographing the aurora in Plate 11-9:

The auroral pictures were made on the way home from an ill-fated kayak sojourn to the Arctic Ocean. I had planned to paddle from Great Slave Lake to the sea by way of the Mackenzie River but was discouraged by an early autumn and by uncertain reports of an anthrax epidemic in the watershed. On the way home to Tennessee I stopped in Kenora, Ontario, for R&R at a posh hotel on Lake of the Woods. From the seventh floor I saw this band of light in the north. For half an hour, I ignored it, thinking "just city lights on cirrus clouds." It took a while for me to realize that there were no cities in that direction no matter how far you went (it's so easy to believe that there are cities *everywhere* that you operate on that assumption even where it isn't true). I drove out into the country east of Kenora and clamped my camera to the open door of my 4WD truck to make these pictures.

Very rarely one can see an auroral display in the southern United States, at times of great disturbances on the sun (after a delay of about 24 hours). Such a display is usually seen only as a red glow in the north, as shown in Plate 11-13 taken from Table Mountain, California. We missed this display because of clouds, but it was seen well from Kitt Peak and the Whipple Observatory.

A great auroral display is awe inspiring. In your mind's eye you can see the fiery Valkyrie of Norse legend sweeping across the heavens in battle array. Most startling of all is the unearthly silence that prevails. Such visual grandeur should surely be accompanied by appropriate sounds, but none can be detected, even by sensitive instruments. Nothing occurring 120 km (75 mi) above the earth generates sound because at that altitude an almost perfect vacuum exists.

NUCLEAR-DETONATION "AURORA"

The ability of energetic electrons and nucleons to produce auroral phenomena was conclusively demonstrated by the detonation of a nuclear device high above the atmosphere. Three times a nuclear warhead was carried up by rocket from Johnson Island in the Pacific, southwest of Hawaii. The first was burst at 43 km (27 mi), the second at 77 km (48 mi), and the last at 420 km (261 mi). The first two were still within the high atmosphere, below the auroral altitude (Figure 9-1), but the last was in the auroral zone. The experiment, on 8 July 1962, carried a 1.4-megaton hydrogen bomb; this shot produced some dramatic effects. Four subsequent shots were smaller and at relatively low altitude. Such experiments were then terminated by international agreement that forbade nuclear explosions in or above the atmosphere.

The 420-km explosion, Starfish, produced some unusual effects. Street lights failed in Hawaii, some 1,290 km (800 mi) away. Burglar alarms and circuit breakers opened. The significance of this was not appreciated for almost two decades; but now there are increasing fears of the effect an atomic burst above a country could have on its power network and its many devices, from radios to gigantic computers, that depend on microcircuitry. What had happened was this: The gamma rays emitted when the uranium or plutonium detonated caused energetic electrons to be knocked out of the atmospheric atoms. These electrons, when captured by the lines of magnetic force, generated an intense electromagnetic pulse (EMP) that induced high transient voltages to appear in electrical conductors hundreds of miles away from the event. Voltage pulses as large as 50,000 per meter of conductor length are possible. Lightning arresters are powerless to intercept such rapid and high pulses, and thus these voltages appear within the circuitry exposed to the EMP. Vacuum tubes are rugged and not seriously affected by transient pulses, but silicon microcircuitry is very sensitive. It is now feared that the detonation of a single hydrogen bomb 500 km (310 mi) above the central United States could affect sensitive systems over the entire country, bringing the nation pretty much to a halt without any physical destruction from a thermal and sound shock wave. Shortly after Starfish was detonated, another consequence became apparent. The communication and scientific satellites that were orbiting the earth began to fail. Solar-cell power dropped steadily and circuitry degraded. What had happened was that these same electrons, and alpha particles as well, had become trapped in the Van Allen radiation belts, augmenting them and adding new ionization belts below the normal Van Allen region. These particles striking the solar cells produced impact defects that raised internal resistance and shortened the lifetimes of the solar-cell electrons, causing the satellites' power systems to fail.

An earlier detonation, Teak, produced some striking visual effects. When the rocket reached the assigned altitude, there was a brilliant starlike flash of the bomb. At once a red bubble of emission expanded from the point of explosion, as shown in Plate 11-14, a photograph taken from Mount Haleakala, Hawaii. There are, in fact, two luminous plasma bubbles, a larger one glowing red from the excitation of atomic oxygen in the upper atmosphere, and a smaller one glowing yellow, perhaps from the diffusion of atomic nuclei from the nuclear event. As the bubbles expand, their light illuminates the clouds lying below the summit of Haleakala and silhouettes the mountain itself. In Plate 11-15 the cloud composed of bomb and rocket debris can be clearly seen, filling the field of view of the camera.

The excitation of atmospheric emissions is quite different from that caused by the higher explosion of Starfish. It was at 420 km, essentially in the vacuum of space where very little atmosphere remains. Much of the material became trapped by the earth's magnetic field, which guided the nuclear debris electrons and ions along its lines of magnetic force until it brought them down into the upper atmosphere, much like particles in the natural aurora. Not surprisingly, the resultant display, shown in Plate 11-16, looks like an aurora complete with greenish-white rays and a background ruddy glow. This photo is one of a spectacular series taken by Herman Hoerlin.

When the atomic bomb exploded, it produced many electrons. These electrons quickly became trapped by the lines of force of the earth's magnetic field. They traveled along these lines, and where the lines returned to the 120-km (75-mi) level they produced faint auroras, both borealis and australis. These man-made auroras, we pray, may never again be seen and these illustrations may be an archival record of an awesome event from years past.

12
City lights

City lights and their glow on the atmosphere above them are also things of beauty. As astronomers, we think less charitable thoughts when they betray their presence on our photographs of distant quasars or in our sensitive measurements. Back in 1955 one of Aden's tasks when searching for the best site on which to build our nation's first national optical astronomical observatory was to find a place where the lights of civilization would not encroach for at least 50 years. A resident back East probably thinks that was a simple assignment. There must be many such places out in the West, where, even today, you can drive for hours along seemingly endless stretches of superhighway without seeing more than a few minimal amenities of civilization – gas station, café, and motel or trailer park. But finding a really dark place isn't that simple. In the clear desert night air you can see the glow of even small communities from afar. In west Texas you can drive toward the lights of a small settlement on a rise across a wide, dark valley, seeing them first as tiny twinkling stars beneath the faint canopy of the night sky, and expect to be there in 10 minutes; but an hour later you are just arriving.

Of all observatories, one of the most excellently sited is the Mc-Donald Observatory, north of the Big Bend country in west Texas, a location selected by Van Biesbroek in the early 1930s. We loved its black skies for astronomy, but it was lonely for Marjorie and our seven children. From our large living room windows looking south forever, you could not in the daytime see Fort Davis, Alpine, or Marfa hidden by foothills; at night they were betrayed sometimes by a trace of glow if there was dust in the valley. You could occasionally see the lights of one or two cars threading their way across the distance.

Not all observatories have been as fortunate as McDonald. In 1902 George Ellery Hale was asked by the Carnegie Institution of Washington to build a solar observatory in the Far West. Yerkes Observatory, which he had built in Wisconsin in 1893, was good but Chicago and Milwaukee were already beginning to cast glows on the skies, though not nearly so strong as the "Chicago aurora" we would see to the southeast at Yerkes 50 years later. Hale liked southern California and the magnificent Sierra Madre to the north of the modest community of Los Angeles. He saw the possibilities of Pasadena's Throop Polytechnic Institute. His colleagues at the University of California joined him in Pasadena, where Robert Millikan transformed Throop into the California Institute of Technology. Hale picked as a site Mount Wilson, rising north of Pasadena, and leased some land there on which to build the solar observatory. There was a 15-km (9-mi) trail, only wide enough for foot traffic in places, leading from Pasadena to the summit. The clarity of the Pacific air made the small clusters of lights 1,741m (5,710 ft) below of small consequence. The site was so excellent that solar telescopes were installed, then a 1.5-m (60-in.) stellar telescope, and later the 2.5-m (100-in.), at that time the world's largest telescope. Marjorie's father and mother had both received their doctoral degrees at Yerkes by 1920, and he joined Hale's growing staff. She could not, even with the distinction of being the first woman to get a Ph.D. in astronomy from the University of Chicago, because there were no facilities for women at the Monastery (dormitory) atop Mount Wilson.

The beautiful climate that attracted Hale to California also attracted many other people. The rapidly increasing population hastened the end of the excellent observing conditions on Mount Wilson. As Los Angeles grew, so did the lights in the valley. Mount Wilson Observatory had earned a fine reputation, as had Cal Tech, so it was inevitable that Hale would want to build a telescope even larger than the 2.5-m. He really wanted a 7.6-m (300-in.) telescope, and his colleagues G. W. Ritchey, the optician, and F. G. Pease, the astronomer, designed one. But being a realist as well as a dreamer, Hale was still quite pleased when the Rockefeller Foundation gave him the go-ahead for a 5.0-m (200-in.) telescope. But where would he place it?

Mount Wilson could not be considered because of the upwelling glow from the cities now spreading across the entire Los Angeles basin. Hale's team began a search for a new site, but insisted it had to be close to Cal Tech and Pasadena where the astronomers had their offices and laboratories. What about a 2-hour driving limit? Discouraging, especially before freeways speeded travel across the basin. There were some sites within a 4-hour drive, however, and one looked good: Mount Palomar, about halfway between Los Angeles and San Diego. So the 5.0-m telescope was installed at Palomar Observatory. But even then astronomers knew that the site's freedom from the glow of cities was limited. Now almost all the coastal regions of California are within sight of the distant glow of some growing community.

Aden's first encounter with the city-glow problem came when he was still a student, about to enter Cal Tech. He was earning tuition money working at the Mount Wilson Observatory optical shop when the United States entered World War II and the lights of Los Angeles were dimmed. The astronomers worked mostly on war-related projects, but when time permitted they scanned the now-dark skies with the 2.5-m telescope. So improved was the view, unswamped by city glow, that even the natural light of the night sky seemed intrusive. If only, the astronomers complained, they could rid the night sky of the airglow. They could, and did. Aden designed an interference filter to block out the chief offending airglow emissions.

When Aden was asked in 1955 to find a new observatory site where future growth of cities and their lights would not produce the problems already plaguing Yerkes, Mount Wilson, Lick, and Palomar, one place he was asked to search was the deserts of the southwest. Their climate was severe, but excellent for an observatory, and at the time the population was small. He did his best, but what constitutes the "best" site? Aden and Helmut Abt explored remote mountains and plateaus; but how would astronomers visiting the national observatory reach it? Where would families live? Where would children go to school? Hale thought living in Pasadena a nearly ideal solution for working at Mount Wilson. But how would one combine convenience and isolation at the same time?

The search narrowed to Arizona's Kitt Peak, a high mountain soaring like an island above the surrounding desert floor. It was near Tucson, a growing city 75 km (45 mi) to the northeast, not directly south as was Los Angeles below Mount Wilson. Astronomers work much of the time in the southern sky, and a city to the south is especially bad. Another thing Aden liked about Kitt Peak was its location in the Papago Indian Reservation. The city could not grow out there. In addition, the Tucson mountains formed a barrier, isolating the broad Avra valley from the city. Even so, would Kitt Peak be safe from city lights for 50 years? Aden didn't think so and, to the consternation of some astronomers on the board, said in his report on the site survey:

The present sky glow from Tucson is appreciable on the NE horizon . . .
The glow does not extend very high in the sky [1958] . . . nevertheless the
situation could become an annoyance if Tucson should expand westward.
The availability of . . . water in the indefinite future could change the entire
face of Arizona, but the development of a space observatory may have ren-
dered this eventuality of minor consequence to the future of astronomy by
that time.[1]

The first artificial satellite, a tiny sphere, had been launched only
months before this statement.

Astronomers still battle city lights encroaching on their terrestrial
observatories. But just how do city glows affect telescopes? Outdoor
lamps shine a large fraction of their light upward and outward above
the horizontal. This light is scattered by the haze that is part of each
urban community. Part of the scattered light is reflected back down,
hiding all but the brightest stars from view. In downtown Los Angeles
the glow is so strong you usually cannot see stars even on a clear night.
In Tucson we still can see the brighter ones, but upon leaving the city
and looking up we can see the sky as star-studded and black. Looking
back toward the city, we see a glowing dome above it, shown in Plate
12-1. If you can see the city lights directly, as from Mount Lemmon
(Plate 12-2), the dazzling sight overwhelms the glow, making it invisible. If you live in the foothills, the sea of city lights below you reaching across the valley is a magnificent sight, especially on a sparkling
clear, cold winter evening. The cold air settling in the valley makes the
myriad lights twinkle like so many diamonds. Even the used-car lots
look magical from this distant viewpoint. But not so to the astronomer – at least not while he or she is at work at the telescope.

From Kitt Peak we can see the Tucson glow to the northwest low
in the sky, but sensitive instruments on the telescopes can already see
the emission from Tucson even in the zenith. Fortunately, digital in-
formation-processing techniques are making it possible to subtract the
city emission from the astronomical emissions. To the southeast the
neon glow of Nogales is largely hidden by a mountain range. To the
west lights are now appearing in growing numbers on the Papago
Reservation, especially at the headquarters at Sells (Plate 10-3). When
Aden first ascended Kitt Peak on horseback in 1956, only three lights
could be seen on the reservation; but when vast copper deposits were
found there village electrification soon followed. One village even has
solar cells powering its modernization.

Two types of light emanate from cities. The first is incandescent
light, a whitish light with a continuous spectrum. The second, and

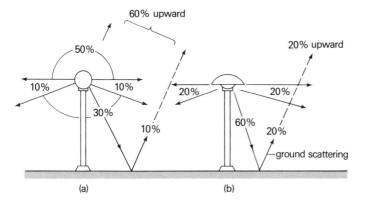

Figure 12-1. Outdoor lamps (a) without and (b) with a shield to control upward light emission. The shielded lamp sheds more light where it is useful and less into the skyglow above a city.

worse from the astronomer's viewpoint, is light from neon, mercury, and sodium lamps. The energy from such lamps tends to concentrate in a few narrow wavelengths; thus although they provide good light for our eyes, they are disruptive to sensitive spectral instruments. Their glow in the skies over the observatory can be invisible to the eye and yet be clearly recorded by the astronomer's instruments, interfering perhaps with some faint galactic signal that has finally arrived after a voyage of billions of years across the vast expanse of the universe. Can you blame the astronomer for being apprehensive that at some time in the near future the interference will be even worse?

Arizona astronomers have been fortunate in gaining local governmental support to minimize the growing intrusion of lights. Shields over lamps help by stopping upward-directed rays, but still 10 to 20 percent of the light illuminating city streets, used-car lots, and outdoor sports arenas bounces back up into the sky. High-pressure sodium-halogen lamps put out little blue and ultraviolet light and thus interfere less than mercury lamps, because much important information is delivered to us from distant stars and galaxies in the blue and ultraviolet. Further, the atmosphere scatters more poorly in the red than in the blue. Already the dominant skylight from Tucson is yellow, as shown in Plate 12-1.

In Figure 12-1 we show schematically the two basic forms of outdoor illumination. The unshielded lamp sends as much light upward as downward. Only 30 percent is directly useful, 20 percent is not really effective because it goes out horizontally, while 50 percent, as well as about 10 percent reflected by the ground, goes into the skyglow above the city. When the lamp is shielded by a white reflective surface, 60 percent of the light is directly useful, 20 percent is ineffective because it still goes horizontally, and only the 20 percent reflected by the ground goes upward to make up the skyglow. The percentage going upward totals 60 for the unshielded lamp compared with only 20 for the shielded lamp – one-third less. This means that a threefold greater population can live as well at night without additional detriment to astronomical observations and at a significant per capita reduction in electrical power requirements for outdoor illumination.

City light on clouds has a beauty of its own, as depicted in Plate 12-3, where Tucson lights are beginning to show on the clouds as late twilight fades. Illuminated clouds can be a dramatic sight from within a city on a rainy night. The lower clouds pick up the glow as they pass over the more brightly lit portions of the city. These low clouds are often fast moving, so the panorama is constantly changing.

Full moon is another occasion when cloudiness combines with light to make the night sky interesting. Full moon even on a clear night severely limits what the astronomer can do. Only the brightest stars can be studied without scattered sunlight affecting the results.

Watching for the glows of hidden cities can be a pleasant pastime on a long drive at night, even a guessing game as to which community you are seeing. The game is best in populous areas and hazy atmospheres when incipient clouds line nearby mountains.

The view out the window of a high-flying jet on a clear night over the midwest can be an incredible sight. If you press your face to the window in order to cut cabin reflections, you can see cities, towns, hamlets, and even individual farms scattered like dewdrops on a spiderweb of roads as far into the distance as the eye can see. Take out a map and play the guessing game. The map takes form below you before your eyes.

13

Twilight on the planets

Space exploration has added a new dimension to the appreciation of sunsets and twilights. We have now been set free from the bonds of earth to visit the planets, circle them, and even land on them for a more leisurely look. Now let us, as did the astronauts, look back at our earth and see it as a planet.

EARTH FROM SPACE

Sunset and sunrise viewed from space are more dramatic in some ways than when viewed from the earth's surface. True, one does not see the entire sky blazing with warm colors, but one does see the awesome contrast between the black of space and the vivid line of color at the limb of the earth. To the astronaut the transition from day to night is very abrupt because the spacecraft is orbiting the earth so fast toward the rising sun and away from the setting sun that sunrise follows sunset in only about 45 minutes.

The blast of sunlight is so intense that it is difficult to obtain good photographs of sunrise or sunset. Several pictures taken from *Skylab* are shown in Plates 13-1 to 13-3. Just before sunrise (Plate 13-1) the upper atmosphere is brilliantly illuminated by sunlight, showing the blue of Rayleigh scattering above and the whiter color from added aerosol scattering below. The lowest layers are still in the earth's shadow, but red light comes through from the sky beyond the haze that lies close to the surface. In Plate 13-2, the upper edge of the rising sun appears, greatly flattened by refraction. In Plate 13-3 the sun has risen enough to create a glare in the camera. Notice also that now the sun lights up the clouds lying close to the surface in a thin bright line at the horizon. A similar blast of light from the flattened rising sun can be seen from balloon altitudes, as shown in Plate 13-4.

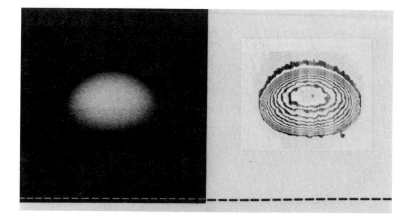

Figure 13-1. The sun photographed from Apollo-Soyuz *when the upper limb was 22 km above the earth and the lower limb 13 km (left). The isophotes (right) show the distribution of brightness over the solar disk. (NASA Johnson Space Center)*

The *Apollo-Soyuz* astronauts used an infrared film and a long-focal-length camera to get a sequence of pictures that show well the sun's distorted flattening, which is caused by the vertical gradient in density of the atmosphere. A photo taken when the upper limb of the sun was 22 km (14 mi) above the tangent line to the earth and the lower limb, 13 km (8 mi), is shown in Figure 13-1. Note the shading of intensity toward the limb of the sun, a phenomenon called limb darkening, which is caused by the solar atmosphere itself. This is not always conspicuous because of the total brilliance of the whole disk.

Moonrise and moonset have provided an easier target for photographing the effects of refraction in the earth's atmosphere. A pair of photographs of moonrise taken from *Skylab* is shown in Plates 13-5 and 13-6. To the astronaut our atmosphere is only a thin band at the edge of the earth. The moon is sufficiently large that its lower edge is in the dense lower atmosphere while its upper edge is almost above the atmosphere. The shape of the moon therefore appears flat on the bottom and close to circular on the top, resulting in the strange ethereal blob rising through the faint blue of the atmosphere. This accentuated refraction is shown in the diagram in Figure 13-2. As the moon rises the flattening decreases until the moon floats as a familiar round object against the deep blackness of space.

At night the astronaut can see the airglow in distinct layers above the limb of the earth, provided cabin illumination is low enough for his eyes to become dark-adapted. The photograph in Plate 13-7 is a

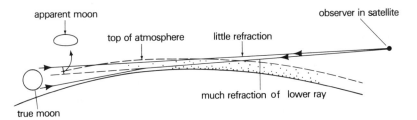

Figure 13-2. Diagram illustrating the anomalous flattening of the rising moon as viewed by an astronaut far above the atmosphere.

time exposure of the horizon. The last remnant of the day shows as a blue line along the horizon. Above this line are two faint layers, the bottom and weaker one reddish and the upper one greenish. The greenish one is obviously from the green line of atomic oxygen, the strongest emission in the visible airglow. The reddish lower one is probably a combination of sodium emission mixed with some OH bands that reach into the red region. One expects the red line of atomic oxygen to arise from a less-distinct region above the green line and not be visible in such a photo because its total intensity is less than that for the green line. In the foreground you see the back of a solar-cell panel of *Skylab*. Moonlight illuminates and casts shadows on the panel.

The three-dimensional appearance of the aurora as seen from orbit is shown in Plates 13-8 and 13-9. This is the aurora australis of 11 September 1973, looping under *Skylab* like a sinuous cloud. Northward of the aurora australis is a diffuse red-colored stratum, a frequent companion to an aurora. Note that this aurora has essentially no color. A red lower border to some auroras is due to its penetration down to where molecular nitrogen becomes excited. The diffuse red glow stratum of some auroras, as in Plate 13-9, is due to excitation of the red oxygen line.

THE MOON

The moon has no atmosphere whatever. The gravity from so small a celestial body is insufficient to retain any atmospheric gases, their thermal kinetic energy being larger than the escape velocity. When the *Apollo* astronauts landed and left the moon, their rocket engines temporarily gave the moon some trace of "atmosphere," but these gases rapidly diffused into space. It therefore was quite a surprise when the pictures returned by the several *Surveyor* spacecraft that safely landed

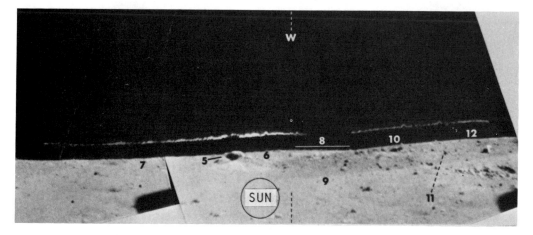

Figure 13-3. Sunlight shining on a lunar dust cloud seen against the darkness of space, from Surveyor 7. The small jiggles, such as near point 10, are transmission glitches. (NASA: J. J. Rinnelson and D. R. Criswell)

on the lunar surface showed something above that surface after sunset – a kind of lunar twilight. One of the clearest pictures of what appears to be suspended lunar dust is shown in Figure 13-3, reported by J. J. Rinnelson and D. R. Criswell.[1] It apparently is a transient phenomenon, where dust kicked up perhaps by micrometeorite impacts, is suspended by electrostatic repulsion. The patchiness of this twilight cloud indicates that it is of limited extent.

The moon is perhaps the ideal place from which to view the zodiacal light (Chapter 10). On earth the atmospheric absorption and the twilight brightness prevent our seeing that part of the zodiacal light very close to the sun. On the moon, as soon as the sun sets below the horizon the full beauty of the inner zodiacal light appears, merging completely into the outer corona of the sun. The *Apollo* crews saw this scene repeatedly, made sketches, and took photographs, one of which is shown in Figure 13-4.

Superimposed upon the general diffuse wedge of zodiacal light the astronauts saw details, streamers, as shown in the sketch reproduced in Figure 13-5. Harrison Schmidt was the first to comment to Houston Control on what he saw:

In lieu of the solar corona photography, I watched – Gene [Cernan] and I both watched, it set and there are two bands which I can still see now – a zodiacal light, I guess, going out symmetrically on either side of the plane of the ecliptic, and they make an angle between themselves of about, let's say, 70 to 80 degrees. I can still – knowing they are there – I can still pick up the bands, streamers, I guess would be a better word. And last night

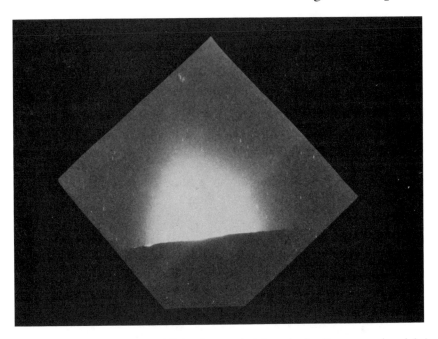

Figure 13-4. The inner zodiacal light photographed from the Apollo *command module in orbit around the moon, taken immediately after the sun set below the edge of the moon. (NASA Johnson Space Center)*

when I watched one set [he is speaking during orbit 61] there was a strong linear streamer going out – of maybe three or four or five – I will have to get my directions straight [they were moving rapidly in orbit around the moon] . . . But those two streamers today are about an equal strength, and they are still visible as zodiacal light.

The next day, on orbit 72, Schmidt remarked: "Just had a good view of the sunset and the corona, and there are two strong bright streamers just right at sunset, one parallel to the plane of the ecliptic and the other – oh, maybe 10 degrees to the south of the plane." The nature of these streamers is a bit of a puzzle: The approach of sunrise would not cause streamers to change rapidly. However, they did become very conspicuous as sunrise approached, as though the sun were shining on particles relatively close to the spacecraft. On *Apollo 17,* Evan remarked that within a few seconds of sunrise the brightening "just zaps out across the sky." They recognized that the *Apollo* capsule does have a cloud of tiny particles associated with it that could have caused the lunar "twilight rays." This was the last of several missions to the moon, considerably stirring the lunar environment. Perhaps now the moon is once more disturbed only by an occasional impact of tiny meteoric particles and the dilute solar wind.

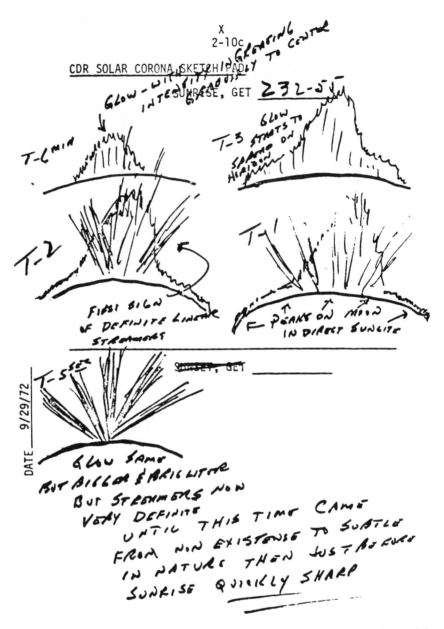

Figure 13-5. Sketches by astronaut Cernan on Apollo 17 *showing how the visibility of the streamers dramatically increases 5 seconds before sunrise. (NASA Johnson Space Center)*

MARTIAN SUNSET

Mars has a very thin atmosphere (about one one-hundredth that of the earth), most of which is carbon dioxide mixed with a little nitrogen. The life-giving molecules, oxygen and water vapor, are barely detectable and are much too sparse to support life. In spite of this thin atmosphere there is something in the air that colors the sky. This coloration becomes quite noticeable as sunset nears (Plate 13-10). There is either some very fine dust suspended by the strong Martian winds or some sort of aerosol, or both. This is evident from the fact that even at noon the Martian sky is reddish rather than the bluish expected on the basis of Rayleigh scattering. While assembling pictures for this book, we found a photo we had taken just before sunset near Nanjing, China, showing the sun seen through a fine veil of dust (Plate 13-11). The dust in these two cases, Mars and China, lies close to the ground, not high like volcanic dust, so that a dust "shadow" is cast on the lower part of the sky in both scenes.

In the Martian sunset scenes you will notice some "contouring" of the colors and brightness. This is purely an artifact of the digital mode of picture transmission. When the scene brightness is low, the discrete levels of the digital quantization show; to get the proper visual impression, you must mentally smooth out the contours.

Notice the size of the sun in the Martian pictures. Mars is 1.5 times farther away from the sun than is earth. You can readily see the position and size of the sun in Plate 13-11 (on the horizon), but this image is slightly overexposed and thus enlarged. Seen from Mars, the surface area of the apparent solar disk is 2.5 times less, as is also the received solar energy. This is one reason Mars is a cold planet; the temperature even at the equator seldom reaches $0°$ C ($32°$ F). The major planets, being much farther away, are even colder. If you were to contemplate going to the nearest stars in search of a planet for a new home, you would be discomforted throughout your voyage by an even more devastating thermal situation. Perhaps engineers could devise machines capable of enduring decades of space travel and even provide for a future generation who would be the ones to circle a new home; but what about the trip itself? The traveling spacecraft would be exposed to the cold of deep space, which, far from any star, would be about $-269°$ C (only $4°$ K). The spacecraft would soon radiate away its internal heat and you would need to find some extremely compact energy supply just to keep your spacecraft from approaching absolute zero for the many decades required to reach the nearest stars. Solar energy or star energy could not do the job. Nothing short of a nuclear power supply would suffice; thus problems mount.

TWILIGHT ON VENUS AND TITAN

On occasion, from earth we can see sunset and twilight effects on Venus. Venus passes between the earth and the sun, sometimes transiting across the solar disk. The next pair of crossings will be on 8 June 2004 and 6 June 2012. The last pair were in 1874 and 1882 and were widely observed to see if Venus had an atmosphere above its clouds. Observers concluded it had because the planet, like a round black teardrop, appeared briefly to hang attached to the solar limb before it broke away and traveled across the disk; the opposite effect occurred as it left the disk.

With proper precautions one can see an effect of the high atmosphere of Venus every time it passes close to the sun, at inferior conjunction. The inclination of Venus's orbit to that of the earth is great enough that Venus infrequently transits the solar disk, yet often appears within a few degrees of the sun. This means that the sun is only a few degrees below the horizon of Venus, and just as on earth, solar rays illuminate the upper atmosphere. When the solar elevation on Venus is small enough, the twilight arch can be seen encircling Venus, as shown in Figure 13-6 (right).

We were delighted recently when *Voyager 2* was able to look back at Titan, the largest satellite of Saturn, and record the same sight we had seen of Venus from earth: a complete ring phase, shown in Figure 13-6 (left). These views are not shown in color because the coloration is very faint. In the case of Venus the ring is bluish and the cusps of the planet silvery white. No reddened light survives passage through the cloudy zone of Venus, and the remaining gas above the clouds is so thin only a trace of bluish scattered light can be seen. Titan shows a bit more color: The planet itself is orange but the ring bluish.

Obtaining this unusual picture of the twilight ring around Titan was a grand adventure in big science – a $500-million space spectacular was involved, but with enormous dividends in advancing our knowledge of Jupiter and Saturn. As of this date this adventure is still continuing, with *Voyager 2* en route to a rendezvous with Uranus and Neptune. Photographing this unusual twilight ring around Venus was also a grand adventure, but in small science. In fact, it was *very* small science. The story is worth relating.

In 1939 four graduate students at Cal Tech decided that the study

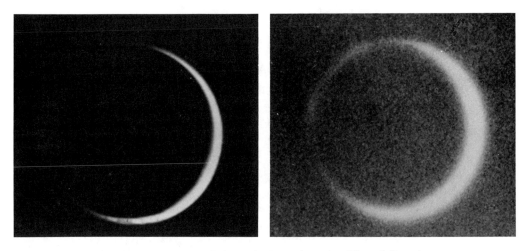

Figure 13-6. Almost identical views of twilight on the Saturnian moon Titan (left) and on Venus (right), showing the twilight arms almost encircling each. If Titan were free, it would rank as a respectable planet. (Titan: NASA, Jet Propulsion Laboratory, California Institute of Technology, Pasadena, 1981. Venus: Ernest Wright, James Edson, and the Planet Group, 1940)

of the planets had been relegated to the back burner by astronomers. James B. Edson, who had worked as an assistant at Lowell Observatory before continuing his graduate work at Cal Tech, decided that something could be learned about the height of the Venusian atmosphere from the observation of the twilight cusps seen when Venus is close to inferior conjunction – when Venus passes between the earth and the sun. The astronomers at Lowell Observatory had observed the passage in daylight, and Edson devised a simpler technique so that even a small telescope could make observations. Edson and several friends – Ernie Wright, Dick Canright, and Jim Winget – decided to set up an observing station at Table Mountain, near Wrightwood, California. Table Mountain then was the site of a Smithsonian Solar Observatory, and is now the site of the Jet Propulsion Laboratory's telescopes. Ernie's father advanced the grand sum of $1,000 to finance the entire project.

For telescopes, they borrowed a 15-cm (6-in.) refractor as a guiding telescope and a 15-cm (6-in.) excellent reflector from Roger Hayward for the photographic telescope. The addition of a small microscope to the reflector enlarged the image so that the disk of Venus was of respectable size on the film. To complete the setup, they constructed a 10.7-m (35-ft)-high steerable boom which had a plywood sunshade at its top. The completed setup as built on Table Mountain is shown in Figure 13-7. Aden entered this adventure shortly after the obser-

Figure 13-7. The refractor-reflector combination used on Table Mountain by the Planet Group to photograph Venus passing by the sun in 1940. The lower portion of the moving tower carrying the sunshield is at the right. (James Winget, James Edson, and Ernest Wright)

vations were made and was given the chore of measuring the photographs. At the same time the group was laying plans for ambitious observations of Mars with a 60-cm (24-in.) telescope they had put together, but the onset of World War II ended these dreams.

Perhaps a view of the Venusian sunset may even be photographed from the surface of Venus. The Soviet *Venera* probes found that about as much sunlight reaches the surface far below the haze layer as reaches Moscow on a June day. The view would be a brief one because Venus's surface temperature of 500° C (932° F) would soon destroy the electronics of any probe. Would the probe see a blue or green sun or only a hazy deep-red spot in the sky? It would be interesting to look back at the sun from a space probe settling slowly downward through the thick atmosphere of Venus.

TWILIGHT ON JUPITER AND IO

Jupiter has an extensive atmosphere that could show dramatic sunset colorations to a properly positioned observer. The strong gravitational pull of Jupiter, however, causes the space probes to move past so rapidly that only a few fleeting glances are possible. Color pictures are made by taking three separate exposures with red, green, and blue filters. The proper superposition and relative intensity of the three then reconstruct the actual color of the scene. The only picture so far obtained of twilight on Jupiter clearly shows the rings in the glare of full sunlight. The separate color frames are poorly registered because of long exposure and rapid motion of the spacecraft. If they were aligned properly the true twilight coloration would appear pale – nothing to compare with what we enjoy on the planet earth.

Looking at the Jovian moon Io, we can see evidence of a transitory atmosphere created by the eruption of gases and vapor from several of Io's active volcanoes. Scattered sunlight makes the volcanic plumes appear blue, as shown in Plate 13-12.

EVENING SKY ON SATURN

Both *Voyager 1* and *Voyager 2* passed Saturn, but no twilight pictures were obtained, inasmuch as all attention was focused on the fascinating unveiling of the details of Saturn's rings. The rings would, in fact, appear as the dominant feature of the twilight and night sky to any inhabitant that could float atop Saturn's icy methane and ammonia clouds. The sun is very far away, 9.5 times farther than it is from the earth, and thus the rings, though faint, would still be dazzling against the black of space.

The drama of space exploration appears to be drawing to a close. *Voyager 2* is scheduled to visit Uranus and Neptune in the years ahead. The spacecraft will be incredibly cold, heated only by its radioactive-isotope power supply. If its cameras still work, we will enjoy more scenes from where man may never go. Perhaps some future civilization on earth will remember these feats of early planetary exploration and resolve to follow the promise suspended tantalizingly before us today.

14

Celestial visitors

"Celestial visitors" carries a feeling of something beyond human inter-vention. In past ages divine intervention was associated with visions seen in the sky. The emperor Constantine saw a flaming cross and the words *en toutoi nikos* (Greek, "in this sign conquer") in the afternoon sky. He obeyed, won the decisive victory the following day, and changed the course of Western civilization. Auroras are intimately en-twined with the Valkyrie riding across the northern skies. Today we tend to lose sight of the deep impression natural phenomena made on people who were in closer contact with the day and night sky than are we modern urban dwellers. Only in the deserts of the midlatitudes or of the icy polar regions can one fully appreciate the range of appear-ances in the skies.

Comets, those pale celestial swords, swayed the course of history in ancient days. An indication of the emotional level that can still be stirred was evidenced by the 1973 appearance of the comet Kohoutek, held to be the herald of doom by some religious groups and predicted to be unusually brilliant by scientists. Instead, it failed to live up either to occult or to scientific predictions as it swung around the sun. At best it could barely be seen as a hazy spot in a very clear dark sky if you knew exactly where to look.

In this chapter we will describe for you a number of different types of celestial visitors – some very rare, some appearing every few years, and others you can see almost every day or night.

COMETS

Some of the most spectacular comets have appeared unheralded, first seen by airline pilots who regularly fly above the haze. In the last few decades these sudden comets have all approached the earth from southern skies, where they had been hidden by the glare of the sun. They have briefly swung northward, where most people live, to be seen in the morning sky before retreating into the depths of space. Rarely have we enjoyed a comet as spectacular as Ikeya-Seki (1965), shown in the morning sky of Tucson as a resplendent bluish sword poised above the Agung sunrise glow stratum (Plate 14-1).

Comets are named after their discoverer(s), a reward for diligence in comet searching. Occasionally a professional astronomer first notices one, but far more often the finder is an amateur astronomer. When a comet is discovered, a telegram is immediately sent to the Harvard Observatory giving the local astronomical (or civil) time of observation, the position of the object in the sky in astronomical coordinates (right ascension and declination), brightness, whether it has a visible tail, and so forth. A tribute to the skill and diligence of Japanese amateurs is the surprising number of new comets they have discovered, often from locations near cities. A pair of binoculars or a 10- to 15-cm (4- to 6-in.) "richest-field" telescope is a necessary tool for the seeker of the usual comets, those that never achieve naked-eye brightness. The rare bright comet that is seen by many people at the same time is not named, but lives out its passage by designation of the year of discovery and alphabetic order of appearance within that year; thus, it can be simply 1947d (fourth comet in 1947).

Comet watching can be frustrating because of cloudy skies, but the reward can be great. David Huestis wrote about trying many times to see Comet West:

I was continually plagued by clouds to my east. I was up each morning at 3:30 A.M. to allow me to drive to an area with an unobstructed eastern horizon. I just didn't think West would be visible above the treeline to the east of my apartment. Thursday morning was clear. As planned, I picked up my girlfriend, who lived next door, and began driving down the highway. All of a sudden she yelled, "There it is!" "It can't be," I said. "We can't see it from here. It's lower to the horizon." "No," she insisted, "It's right up there. Look!" So I looked up to where she was pointing through the windshield. I slammed on the brakes. Sure enough. There was Comet West sparkling like a jewel in a dark predawn sky. A cosmic teardrop. What a beautiful sight. It really didn't look real. Fantastic! Good thing no one had been driving behind me. Here I was stopped in the middle of the highway looking up into the sky. I then pulled across the median strip and rushed home. I took slides of Comet West until dawn's early light blotted it all out.

A comet is most probably a loosely bound collection of dirty ices, remnants of the material from which they were formed, and whose home lies far beyond the most distant planet, Pluto. Thousands of years pass before the return of most of these suddenly appearing bright comets. None appears to be a visitor from beyond the fringes of the gravity field of our sun. If they were, they would pass the sun in hyperbolic orbit, never to return.

When a comet nears the sun, its material is warmed and the ices – mainly of water, ammonia, and methane – vaporize and are caused to glow by the excitation of the solar ultraviolet light. Tiny solid particles of dust also are ejected, perhaps the way popcorn is popped, by tiny explosions of vaporizing gas. Both the gas and dust are driven off by the pressure of sunlight to form the most distinctive feature of a bright comet – the tail. Because the tail is driven by sunlight, it always points away from the sun. When the comet swings rapidly around the sun, as a sun grazer, the tail is curved because material leaves the comet at different times, each portion moving directly away from the sun along its own path. The photograph of Comet West in Plate 14-2 shows the spreading out of the whitish dust tail as the comet swings around the sun in its parabolic orbit. It also shows the bluish gas tail pointed more directly away from the sun. This comet still has a conspicuous head or nucleus, as contrasted with Comet Ikeya-Seki, whose head was essentially disrupted by close passage to the sun.

The sun-grazer comet (Plate 14-1) can be especially spectacular. As the name implies, such a comet sweeps very close to the sun. Solar heating is intense – enough to cause even iron to be vaporized. The daylight comet Ikeya-Seki was a sun grazer; it became so bright it could be studied at noon with the solar telescope on Kitt Peak. The gravitational pull of the sun can become so great that the friable conglomeration of the head breaks apart and the comet leaves the sun as several smaller comets.

A comet may pass close enough to a major planet, such as Jupiter or Saturn, so that its orbit is significantly altered by the gravitational pull of the planet. If the pull adds enough velocity to the comet, it can leave the sun never to return. If the pull subtracts velocity, the orbit is lessened and the comet returns sooner. Halley's comet, one of the great ones, returns after 75 ± 1 years, losing more gases and dust each time until eventually all that remains will be the desiccated rocky material. Even this material slowly distributes along the orbital path and will appear some night far in the future as a meteor shower each time the earth crosses its orbit. Halley's comet must be a relatively new member of the periodic comets because it still gives a spectacular show whenever it and the earth are favorably placed with respect to each other and the sun. "Young" is a relative term. Halley's comet is young in relation to the age of the planets, but old in years. Passages of Halley's comet are now identified almost as far back as 240 B.C. There are other periodic comets of far greater age, now identified only by showers of meteors, such as the great Leonid shower of 1833. Biela's comet, on the other hand, still can be seen on passages as a comet even though it also caused intense meteor showers in 1872 and 1885 when the earth crossed its orbit within 100 days of the comet itself.

VAPOR TRAILS

The jet trail or contrail made by high-flying jet aircraft is a celestial visitor that should doubly impress us because it is not only a beautiful natural phenomenon but also a work of mankind. Yet it loses impact because it is so common. Imagine, if you will, a contrail appearing without warning to people who have never seen one and know not from whence it comes. A comet sits serenely in the evening or morning sky, changing position only with the passage of days. A meteor dashes briefly, disappearing seconds after a fiery ride. But a contrail moves majestically across the sky, allowing time for human appreciation. Our forefathers never viewed this sight, and perhaps the day will come when our descendants will no longer see it. Will an energy famine associated with depleted supplies of fossil fuels stay the wings of these high-flying chariots?

In Plate 14-3 we show a typical contrail that was formed by a transcontinental jetliner passing over Tucson near sunset. Because the upper atmosphere is very cold – usually about $-60°$ C ($-70°$ F) – only a tiny amount of water vapor is sufficient to saturate it. The jet engines inject a wake of invisible water vapor along with tiny combustion nuclei on which the cooled water-vapor wake condenses, forming a visible cloud track. Sometimes these contrails quickly evaporate, as in Plate 14-3. At other times, when the ambient air is more nearly saturated, the contrails persist, augmented by others, until the sky is draped with filmy cirrus. At sunset these man-made clouds take on the beauties of the fiery skies. They seldom are seen at sunrise because the daily sequence of jet flights has not yet commenced and the trails of yesterday have evaporated, leaving a clear sky.

SPACE SATELLITES

A few years ago we enjoyed another type of man-made celestial visitor. Thousands are now in orbit, but few have been bright enough to be really conspicuous. These early visitors were named *Echo 1* and *Echo 2*. On many evenings we would look up as twilight faded to see a starry light moving among the stars. The *Echo* satellites are no longer there, yet for several years they visited regularly. Each was a balloon of silvered plastic, about 30 m (100 ft) in diameter. They were tenuous objects designed to measure the traces of atmosphere at their orbital altitude, initially about 500 km (300 mi), gradually coming lower as atmospheric drag drained their kinetic energy. They were highly visible objects on each pass, rivaling the constellation stars, until they winked out upon entering the earth's shadow.

Space objects – real space objects, not elusive UFOs – are still orbiting the earth, but they are smaller and fainter. *Skylab* plummeted to a fiery death over Australia, but *Soyuz* still orbits about 250 km (170 mi) high. A careful observer can see the embryonic space station in the twilight sky, but it crosses the sky in a few minutes so the chance is small of seeing it without the help of a scientific timetable.

Some people dream of the day when man-made "stars" will be seen each night, the reflection of sunlight from permanent space stations orbiting the earth. There even may be colonies supporting space-based manufacturing activities. Imagine a captured asteroid being carefully consumed by such a colony for ultimate delivery either to earth or to projects still deeper in space. Better yet, imagine a solution to our inevitable energy crisis in solar-power farms many kilometers square, capturing and delivering power to a needy earth.

METEORS

Although comets visible to the naked eye are rare, only a few per decade, we see their smaller companions nightly as meteors. Meteors move swiftly, and usually are visible only for a second or two, as shown in Figure 14-1, before burning up as they plunge into the earth's upper atmosphere. Diligence is required of the observer seeking a meteor. If you do see one, by the time you can alert a nearby friend the meteor will most likely be gone. Several trails from small meteors are shown in Figure 14-2, where the camera was pointed at the radiant of the Leonid shower, the place in the sky from whence they appear to come.

Occasionally a great meteor streaks across the entire sky, disappearing beyond the horizon. The path lasts for tens of seconds because these meteors graze the atmosphere instead of plunging directly in. These objects are termed fireballs because of their great brilliance. They

Figure 14-1. A shower of Leonid meteors, 17 November 1966, as seen from Kitt Peak, showing a burst of brightness near the end of their flight. (Donald Pearson)

Figure 14-2. Photograph with camera pointed at the radiant in Leo during the 17 November 1966 shower, as seen from Kitt Peak. (Dennis Milon)

can be as bright as the full moon, a dazzling sight at night; they can be visible even in full sunlight. J. O. Langford related the visitation of such a celestial wanderer over Texas in the 1880s:

It was that fall, not long after my school had started in September, that the meteor fell. At least, I suppose it was a meteor, although I'll probably never know with complete certainty just what terrible phenomenon of nature caused us that awful fright in the night.

The weather was still hot. I never liked to sleep in the house during hot weather; so I had a bedstead and mattress under the river-cane arbor on the east side of the house. We always got a good cool breeze here, and almost never did the mosquitoes come that high up out of the Rio Grande Canyon.

On this night, Lovie and I were sleeping outside; Bessie and the baby slept in the house. It was sometime after midnight that I started awake to find the world about us illuminated by a great white light, and the air filled with an awful cracking and rumbling, as if whole mountains were breaking apart and tumbling down.

Without even thinking what I was doing, I leaped out of bed and ran out into the open yard. Behind me, I heard Lovie cry out. "Daddy!" she screamed and came running to cling to me. And then here came Bessie, clutching the baby in her arms, her frightened face plainly visible in the great light.

"Oscar!" she cried. "What is it, Oscar?"

And of course, I couldn't answer her. I didn't know what it was and was probably too terrified to talk if I'd had the answer. All I could do was hold Lovie in one arm and pat Bessie on the shoulder with the other hand and stare out at a world that was surely coming to an end.

Cowering against the house, Tex lifted her nose and howled in a way that made cold shivers run up my spine. I was conscious, too, of the rooster crowing and crowing again, and from out at the little corral I'd made for Boomer came the sounds of his frightened snorting and lunging, then finally a wild braying.

How long the light and sound lasted, I don't know. Evidently, for half a minute, at least. For I recollect looking out over the great gash of the Rio Grande and seeing in minute detail every rock and crevice and ripple on the water and every hill and bush and rocky ravine, clear to the blazing face of the Carmen Mountains. It was the same when I looked south toward the San Vicente Mountains. Every feature of the land stood out in bold detail, illuminated by a light that was brighter and more glaring than a noonday sun.

The light and sound seemingly had no source; it was just there, all about us, making us quake with fear.

Then suddenly, there was a greater sound, like some sort of terrific explosion. The light flared brighter, reached an almost blinding intensity, then went out, leaving us in total darkness. While we stood blinking in the dark, successive waves of sound slammed up against the Carmens, rolled back to go rumbling up the Rio Grande till they struck the San Vicente Mountains, then returned down the Rio Grande, slamming and banging against the canyon walls again.

It all happened so quickly and was so terrifying that, when the sun rose the next morning, Bessie and I could hardly believe it had happened. It was more like some hideous nightmare that you want to forget as quickly as possible.

But nightmarish as it had seemed, there'd been more substance to it than a bad dream. José Díaz, a rancher from across the river in Mexico, had seen it, and came down to the house the next morning, still alarmed.

"It is a fact, Señor," he said in a voice of awe, "that the very earth shook beneath my feet!"

Later, I learned that others had heard the sound and seen the light clear beyond Persimmon Gap in the Santiago Range forty miles to the north.

Some fifteen years later, I ran across the only explanation I ever found for the thing that happened that night. I was exploring some Indian campgrounds in the foothills of the Carmens for artifacts, and my guide, Juan Luna, led me to a great hole in the earth some five or six miles from Boquillas. It was a great bowl-shaped depression, something like a hundred feet in diameter and possibly half that deep, with pulverized dirt and broken rock rimming the edges.

If it was a meteor that made that terrible light and sound that night, then I'm convinced that the crater Juan Luna showed me is its resting place.[1]

Noise accompanies great meteors that fall to earth. Their entry shockwave reverberates across the land. The time between visual sighting and arrival of the sound tells how far away the object passed. Like the lightning–thunder relationship, each 5-second delay equals 1.6 km (1 mi). Tales of noise simultaneous with sighting have no scientific rationale and must be ascribed to memory compression effects, except when the meteor passes very close, as was the case of the Boquillas meteor.

In 1972 the earth was grazed by a very large meteoroid, and the thought of what could have happened should send chills through everyone. The object was first seen over Kansas in a bright blue sky. At first people thought it was an airplane on fire, but when it rapidly disappeared northward with steadily growing brightness it was recognized as a rare daytime meteor. Over the Dakotas the object's passage was followed a few minutes later by loud thunder as the sound wave reached the earth's surface. The object proceeded on and was last seen over southern Canada, no longer accompanied by the trail of thunder. The object then receded into space, into a new orbit as a consequence of the near collision with the earth. It will pass near the earth at some future date because its new orbit will pass close to the point in space where the orbit became established; the meteor may or may not come so close again.

There was considerable interest in calculating how massive was this visitor that just grazed the earth. The deceleration of the object due to the resistance of the atmosphere gives one way to estimate its mass. Because it traveled an airpath of several thousand kilometers with only a small reduction in velocity, it must have been very massive. Combining brightness (estimated from many pictures) with the deceleration estimate, scientists think it could have been as heavy as 1,000 tons. Thus it was about 6 m (20 ft) in diameter, very tiny as asteroids go, but large enough to have made quite a respectable crater if it had impacted. By comparison, the largest meteorite, found in South Africa, weighs only about 70 tons. The remarkable thing was that this visitor did not break up. Usually the dynamic pressure of the air breaks up a meteor into a shower of fragments, of which few reach the ground. Perhaps this massive visitor was an iron meteoroid.

Since 1958 the earth has had a new source of "meteors." Man-made space junk is continually reentering the atmosphere. Because over 2,000 separate spacecraft have been orbited by the Soviet Union since then, and because each launch in addition contributes many small pieces, it is possible that if you do see a meteor it is actually man-made. There are two ways to tell the difference. First, natural meteors travel at high velocity, 15 to 70 km (9.3 to 43 mi) per second. Those hitting the earth head-on are, of course, traveling the fastest. Reentering space junk travels much more slowly, perhaps only 4 to 8 km (2.5 to 5 mi) per second and often breaks into smaller fragments that may glow green or blue as their magnesium and aluminum burns.

The second way to recognize space junk is by its direction. All satellites travel either from west to east (never east to west) or, especially the military surveillance satellites, north to south or south to north (polar).

EARTH CROSSERS

Astronomers have discovered about 50 small asteroids that cross the orbit of the earth and estimate that perhaps a thousand objects large enough to produce a significant impact crater on the earth still are in orbit. Meteor Crater in northern Arizona is a relatively recent reminder that big things still hit the earth. Erosion quickly removes evidence of such meteor impacts, so those that occurred several million years ago have left no trace except for subtle evidence from shatter cones impressed on basement rock.

A dramatic moment for astronomers came when the first pictures returned from Mars showed the planet pocked with craters. Scientists confronted with these returning scenes at first cautioned against believing what the eye was seeing. But the growing evidence of more and more pictures transmitted to earth showed beyond doubt that Mars was heavily cratered. We knew that our moon was heavily cratered from the infall of planetary debris during the short epoch immediately after the planets and moon were formed. Evidence from this early day has been obliterated from earth by erosion processes; thus the disappointment of seeing craters on Mars. The lack of erosion processes reduced markedly the chance that Mars was a safe haven for life. We now have pictorial evidence that Mercury is cratered even more heavily than the moon, and that the satellites of Mars, Jupiter, and Saturn also bear the scars of ancient bombardments.

Are these bombardments over? Did all the available planetary debris rain down in the first million years? The answer to both questions is no. A few objects – but only a few – remain that are capable of striking the earth, which is not surprising when you consider that the earth has had 4 billion to 5 billion years (American billions) to sweep its orbit clean. It is likely that today's thousand or so earth crossers were perturbed from interplanetary regions into their present orbits. It is an understandably uncertain estimate that one of these small asteroids will strike the earth once each 50 million to 100 million years. And it is therefore interesting to note that perhaps the last impact of an object in the 10- to 20-km (6- to 12-mi) diameter class was apparently 70 million years ago.

What are the possible consequences should earth greet such a celestial visitor again? The meteorite that excavated Meteor Crater, Arizona, some 22,000 years ago was less than 0.1 km in diameter. The cratering by an object of over 1 million times greater mass would throw vast quantities of debris into the atmosphere. The amount injected by such volcanoes as Krakatoa and Santorin is insignificant by comparison.

If a 10-km asteroid should strike the earth, it would make little difference whether it landed in the ocean or on land. The debris would fill the air and rapidly bring night to the entire earth. It conceivably could block plant growth for several years, with disastrous consequences for all life that had escaped the initial blast effects.

That there could have been this type of impact 70 million years ago has been deduced from a layer of debris deposited in limestone layers, debris containing an unusual abundance of the isotope of the platinum-group metal iridium. This abundance is characteristic of meteoric material, but not of natural ores. The date also coincides with the abrupt end of the age of the great dinosaurs, a period known as the Cretaceous-Tertiary catastrophe because thousands of abundant species disappeared forever. Did the darkness following this asteroid event extinguish plant life, and along with it, animals that depended on this source of food? Some scientists think so.

Astronomers are interested in discovering as many of the earth crossers as possible and soon hope to have specialized telescopes to search out these asteroids. Once a definitive orbit is determined, the future position of the asteroid over several hundred years, as well as the associated gravitational perturbation, can be calculated. Scientists can then estimate its potential threat. If a threat is detected, it is now within the power of mankind to avert disaster by shifting the asteroid into a safe orbit. A space vehicle could visit the asteroid on a close approach years before the dangerous time, plant a nuclear explosive, and detonate it when the asteroid had achieved the proper orientation, nudging it into a new orbit.

When the 1972 earth grazer entered the fringes of the atmosphere, it became hot enough to be noticed by the infrared missile-detection satellites as an object streaking in over many states and moving out again into space. This raises the specter that a celestial visitor may be mistaken for a hostile missile. If the asteroid had struck either the United States or Russia, might the immediate response have been the unleashing of a nuclear disaster?

NOVAS AND SUPERNOVAS

In addition to comets and meteors, other celestial visitors appear in the night skies as temporary stars called novas. Some flare up from invisibility to take temporary place among the bright stars of the constellations, then after some weeks fade from view, disappearing to the faintness from which they arose.

Comets are readily recognized because they are fuzzy objects, often with tails. Novas and their more intense cousins, supernovas, look just like any other star, simply points of light. To recognize a nova requires knowledge of the patterns and brightnesses of the visible stars. As some 4,800 stars can be seen by the unaided eye, it is no wonder many novas are discovered only by accident, after the fact, by comparing photographs. Even a bright nova can go unnoticed by professional astronomers.

Early one November morning in 1942, Edison Pettit, Marjorie's astronomer father, went out to pick up his morning newspaper in Pasadena, California. As he had done all his life, he looked up at the sky. There in the south was dawn's soft light, but something looked different that morning because there was a star near the southern horizon outshining the dawn. He couldn't see much of the southern sky, just a strip visible between trees. He ran to his small backyard observatory, rolled back the roof, and pointed his 15-cm (6-in.) Alvin Clark refracting telescope at the new object. With a visual photometer he measured its brightness and position. It was a nova – Nova Puppis 1942 – and a very bright one. Dr. Pettit was to follow the light history of this object for over a year, until it faded away.

Notifying Harvard Observatory, he learned he was not the first discoverer, priority going to an observer in the southern hemisphere. Novas are not named as comets are, but Dr. Pettit was pleased that he was the first to see it in the northern hemisphere. Some other astronomers had good reason to feel embarrassed. On Mount Wilson that morning no astronomer had noticed the nova, a remarkable fact since the pathway from the 2.5-m (100-in.) and 1.5-m (60-in.) telescopes to the Monastery runs due south, and the nova blazed forth in undimmed glory right down that path.

That same night down in Altadena another astronomer, Dr. Dorothy Davis (Locanthi), had set up her camera to photograph star trails. She took four exposures, which made the trails look like the Morse code symbol . . . — , the V-for-victory slogan of World War II. She unwittingly had photographed Nova Puppis, the brightest star in Figure 14-3. Dr. Davis entered the picture in an Eastman Kodak Company contest and won. She writes, "You may be interested in a consequence of selling that picture: It led indirectly to meeting my husband. Bart happened to be in the radio store when I went there to get a transformer for the radio I bought with EKCo's payment."

If you want to discover your own star, a good star atlas, such as the *Norton Atlas,* is indispensable. In addition, you must memorize the constellations. Most novas occur close to the plane of the Milky Way,

Figure 14-3. Nova Puppis (position shown by the arrows) as photographed by accident on the morning of discovery, 1942, when it was the second-brightest star in the winter sky. (Dorothy Davis Locanthi)

which somewhat simplifies the task by reducing the number of constellations to be memorized. The very bright nova, meaning one close to the solar vicinity of our galaxy, can occur anywhere in the sky.

Ancient records often recorded celestial visitors, not for their own sakes but as portents of things to come. One of the games astronomers have played over the years is uncovering these ancient records and then searching to see if some remnant can be detected today. A notable object appeared in the sky in 1054, as recorded by Chinese astrologers. The position they recorded corresponds to that of an unusual object today, the Crab nebula. Measurements of expansion of the gaseous filaments extrapolated backward show the filaments all originated from one spot at about the year 1054! The Crab nebula must be the remains of a millennium-old nova.

Renaissance astronomers also noted new stars. The Danish astronomer Tycho Brahe in 1572 noted a bright new star in the constellation Cassiopeia, visible even in daylight. This caused fears among the people of that day because they believed it portended some great event. Only 32 years later, in October 1604, another extremely bright star appeared in the constellation Ophiuchus, outshining Jupiter. It was witnessed by another great astronomer, Johannes Kepler. Both Tycho's star and Kepler's star left remnants, but they weren't as easy to find as those of 1054.

Remnants can be nothing more than wisps of nebulosity. The astronomer Fritz Zwicky searched for them in 1941. He had a new 20-cm (8-in.) f/1 Schmidt camera, a portable design by Russell Porter of 5.0-m (200-in.) telescope fame. Zwicky wanted to try some deep-sky photos in the light of hydrogen alpha emission, which is the best for detecting faint gas filaments. He wanted to go to the High Sierra for the darkest, clearest skies and needed a husky young assistant. Aden, then an assistant optician at the Mount Wilson Observatory and a Cal Tech student, was drafted for the job. The place they chose from which to observe was Saddlebag Lake near Tioga Pass, a beautiful timberline spot. The skies were dark – almost: Low on the northern horizon above the granite Sierra peaks glowed an aurora! Zwicky was furious because a storm was expected the next night and he and Aden had to leave to avoid being snowed in at above 3,050 m (10,000 ft) elevation. The sky photos were excellent, but Kepler's and Tycho's stars remained undetected until years later when their remnants were spotted by a different tool – the radio telescope. Thirteen years later we drove up to Saddlebag Lake en route to Lick Observatory from Yerkes Observatory. Aden wanted to talk to the Lick astronomers about the possibility of building a 10-m (400-in.) telescope. Today, almost 30 years later, interest in very large telescopes is renewed. Perhaps this time a 10-m will become a reality.

Each Christmas season the media turn renewed attention to the Christmas star. What was it? Did one of these celestial visitors herald the birth of Christ? Was this biblical event a temporary star, a conjunction of planets, or an expression of what lay deep in the hearts of the early disciples as they wrote about the drama of events of their youth when they walked the dusty roads of the land of the patriarchs with Jesus, the Christ? A new star had arisen in the House of David and had shone over the place where Jesus lay. Why do we still wonder about the meaning of that star and search in vain for a physical remnant to prove or disprove what never was meant to engage our minds outside the spiritual domain? The celestial visitor at that first Christmas was for our hearts, not for our telescopes.

15

Reflections

Pursuit of sunsets, especially of volcanic ones, has led us to wonder what sunsets were like when the earth was young. Was volcanism ever so continuous that it changed the climates? It is only natural to think that the atmosphere has always been the way we see it today; the seas and continents also. We now know, however, that the continents have short lifetimes before significant changes occur – short, that is, when the unit of time is 1 million years. The seas must continually grow more saline because salts are transported thence by rivers, but the water returns pure as rain over the land. What about the atmosphere?

The atmosphere is a remarkable heat engine, rather well stabilized against change. Even so, climate is variable depending on where you live. We live in a desert climate and enjoy magnificent sunsets. The low humidity of these latitudes gives hot clear days and cool clear nights. The atmosphere has little thermal inertia. If you live in the tropics the situation is far different. High humidity holds heat in the air at night. The range of temperature from day to night is in the order of 20° C (36° F) in the dry season on the desert, but only 2° C (4° F) in the equatorial rainy regions. Sunsets tend to be crisp and colorful in the desert and soft and pastel in the tropics and cloudy latitudes.

Every time there is a large volcanic eruption the media speculate about whether this one will cause weather changes. Historically, large eruptions appear to have precipitated weather consequences in the years immediately following the explosion. In April 1815 Tambora erupted in the Dutch East Indies in what was probably the largest injection of ash and gas into the atmosphere in recorded history. One year later fogs obscured the sun all summer, chilling Europe and America with "the year without a summer." New England farmers referred to 1816 as "eighteen-hundred-and-froze-to-death." Crops failed to ripen, and famine gripped Europe. When winter came, it was long and severe.

Was this chance? Or did the veiling of the sun so reduce the solar input that the whole earth was cooled? Tambora was south of the equator; yet northern latitudes suffered.

Still earlier, when a volcano erupted in 1783, the northern hemisphere was blanketed by a lasting dry fog. Benjamin Franklin, curious about many things, wondered whether this was the cause of the severe winter of 1784. One occurrence could be chance; two are less likely to be accidental. The volcanism–weather relation is still a question. The climate goes through such large swings that unusual years now and then are to be expected. The glaciers of Europe, waning today, were advancing between 1500 and 1800. Polar seas froze, and tenuous civilization in Greenland was terminated because no supply ships could penetrate the Arctic ice to reach the starving population. The warming trend of the past 20 millennia then resumed, but with fluctuations. What was the situation in ages past when volcanic activity may have been far greater than today?

Volcanoes eject ash in large quantities along with two important gases: carbon dioxide (CO_2) and sulfur dioxide (SO_2). The ash produces dry fogs and blocks incoming sunlight, reflecting some back into space and absorbing some. Is the net energy absorbed in the earth reduced or increased? (Certainly energy now is absorbed in the atmosphere, not at the surface.) Exactly what the ash veil does is still debated, and many computer models have been written in an attempt to answer this question.

Two observational facts led us to the conclusion that the atmosphere of the young planet earth was very different from what it is today. The first evidence comes from the existence of thick strata of carbonate rock, with sometimes thick beds of coal lying between them. The carbon dioxide bound in that rock had to have been once in the atmosphere before the seas fixed it in sedimentary mud layers. A simple calculation shows that if all that carbon dioxide were in the atmosphere the surface pressure would be about 80 times as dense and the temperature consequently much hotter. Venus has a carbon dioxide atmosphere of about this density and a surface temperature of about $400°$ C ($750°$ F).

Why the big difference in present temperatures between the earth and Venus? The two planets are twins with regard to size. The difference may lie in the nature of the material out of which the planets were formed. Hydrated rock loses its water of crystallization at a lower temperature than its carbon dioxide. Material at the distance of Venus from the sun could have lost its water before the material was gathered

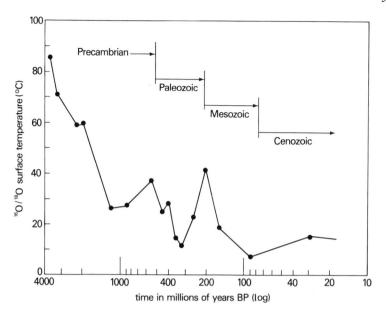

Figure 15-1. Paleo temperatures as indicated by the ratio of oxygen-16 to oxygen-18 derived from dated nodular and bedded cherts, plotted on a logarithmic time scale that compresses ancient times. These calculations provide strong evidence that earth temperatures were much higher in ages past than at present. (Data from L. P. Knauth and D. R. Lowe. Earth and Planetary Science Letters 41, *209, 1978)*

into the planet. The earth, on the other hand, being farther away, held water bound to the material and it was gathered into the planet. The earth, possessing water, was able to fix the carbon dioxide into rock, so that a thin atmosphere remained. Venus, without water, could not thus remove the carbon dioxide from its atmosphere and remained much as we see it today.

The second observational fact is that the ratio of the isotopes oxygen-16 to oxygen-18, as measured in dated nodular and chert layers, shows that temperatures were much higher in ages past (Figure 15-1). In this figure we see a steady decrease in temperature with time, and upon the trend we see rather large fluctuations. We note that the temperature at the present time is about as cold as it ever has been, not much above the thermal minimum, which occurred some 80 million years ago, about the time when the very large animals disappeared from the earth. Astronomers know from stellar model calculations that this temperature history is not the result of the sun's being hotter during the time span of Figure 15-1. The only interpretation is that the early atmosphere was much denser with molecules that also increased the greenhouse effect.

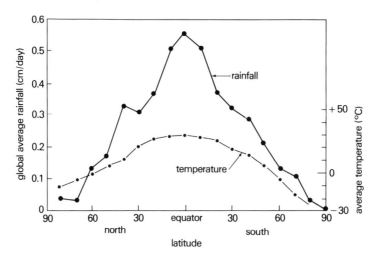

Figure 15-2. Variation of average global temperature and rainfall with latitude.

Surface temperature has a major effect on climate. As temperatures rise, more water vapor is carried by the air. Cloudiness becomes universal, and rainfall reaches very large values. Temperature swings from day to night do not occur; even seasonal swings are lessened. The earth 200 million years ago would have been hot and rainy, with stable temperatures. There would have been no noticeable sunsets, only a transition from daytime gloom to nightime darkness.

A relationship between temperature and rainfall can be seen on the earth as a variation with latitude, as shown in Figure 15-2. The relationship between the two is almost exactly that between the temperature of air and the amount of water the atmosphere can hold. If we apply this relationship and integrate the water content with altitude, we find that rainfall at the thermal maximum of the Mesozoic age 200 million years ago could have been 50 times greater than today. This amount must have led to a drastically different climate and biosphere than today's, as evidenced by the laying down of vast coal deposits.

The high-temperature epochs are interesting to speculate about. This ancient hyperthermal climate would have been very wet, and certainly perpetually cloudy. Two results would have had great impact on the type of animal life. First, temperature fluctuations would have been greatly diminished relative to today because of the thermal control. Second, the atmosphere would have resembled the shallow seas that covered the southern midwestern United States, blurring the difference between land and aquatic environments. Fossil remains testify to the existence of great animals, some clearly aquatic and others, such as the massive brontosaurus, who would have appreciated a swamp to help support their great bulk. The land and sea were intermixed, an ideal setting for the drama of life; but the setting also held the potential for disaster for many species until each had adapted to living on land or in the sea.

The seas were inexorably reducing the carbon dioxide into growing deposits of limestone. When the carbon dioxide was reduced to a certain level and temperatures decreased, a new phenomenon occurred within a relatively short time. The skies cleared, the sun shone through, and temperature swings began. Those animals able to cope with this change and whose means of reproduction were shielded from these temperature swings survived. Those unable to adapt perished rapidly. Those in the seas were protected by its thermal inertia and faced a lesser crisis.

People who live in a cloudy climate such as that in the Pacific Northwest or northern Europe experience in winter what may have been typical of much of the earth in earlier epochs: scarcely any drama in the passage from day to night and night to day. The sun, hidden by a thick layer of clouds, has no real opportunity to announce its comings and goings. We have spent most of our lives on the desert, but during our college days we lived in the fogs of the San Francisco Bay area. How nice it was each time we had an opportunity to spend a few days at Lick Observatory up in the bright sunshine above the fog!

One December we went to Finland to lecture on optics and solar energy. It was night as our plane dropped below the clouds for the approach to Helsinki. From our window we could see sinuous lines of yellow sodium lamps lighting up country roads, yet no city was nearby. When Helsinki came into view, how brilliantly and beautifully it was lighted! Over dinner one evening in a restaurant atop a tower, with a view out over the city, our host remarked he was glad the snows had come, whitening and brightening the land. He explained that in late fall after the clouds have settled over northern Europe, the atmosphere becomes oppressively dull in spite of the ever-present sodium lights. We remarked that this cloudiness is their blessing: The warm sea air, a legacy of the Gulf Stream, shields their northern latitudes from losing heat to a clear sky and gives them their relatively mild winter weather. He agreed, but explained that the pervasive gloom took an emotional toll: In Finland the suicide rate peaks in the dark part of winter.

Our days in Finland gave us desert dwellers some thoughts to ponder. What little gloomy daylight there was at noon only gradually developed after midmorning. Not long after noon the dimming began, foretelling the all-too-early sunset. Up in our hotel room we listened to the radio. Russia dominated the airwaves with stations across the bay in Leningrad. The mournful sound of much of the Russian music played in the darkening midwinter days struck a sympathetic chord; in it we heard a reflection of the soulful wish for the return of the long bright days of summer. When we left Helsinki, we settled back in our seats as the plane rose up into the cloud layer. On breaking through, the sight was spectacular. There on the southern horizon shone the sun, its rays piercing the cabin windows onto the opposite wall. Everybody on the plane broke into a spontaneous cheer! We joined in, now fully appreciating how much the sun means to people who spend much of the winter without these precious rays.

Today's temperatures are close to the historical minimum, but we are free from the even more severe climate of the Ice Ages. Will temperatures rise? It is possible that the drift of continents could cause such a change, if subduction of some of the vast carbonate rock strata leads to intense volcanism and the release of this carbon dioxide. The burning of fossil carbon in our industrial age could trigger a small increase – small in terms of geologic history, but with major consequences for human activities.

Nevertheless, viewed in the light of ages past the present increase of carbon dioxide in the atmosphere poses a smaller threat. We will survive, as we did in the Ice Ages, adapting as necessary. The future is not the same as the past; change is as inevitable as season following season. We live precariously, depleting the fossil fuels produced in more favorable ages to sustain us through winter's chill or summer's heat and to fuel our civilization. Our legacy from ancient times has given us the leisure to pursue art and science and to think about the past and future.

We hope this book has given the reader a renewed appreciation of the glorious sunsets, twilights, and evening skies of, especially, the desert. The earth was not always thus and may not be so at some far future time. Day once passed to night and night to day without the visible presence of the sun, moon, and stars. But what a glorious time it must have been when in the firmament above they, too, appeared to human eyes. We are indeed blessed with a beautiful abode – our Earth, the best of all the planets.

O God of our salvation . . . thou dost make the dawn and the sunset shout for joy!

Psalm 65:5,8
(c. 1000 B.C.)

Notes

CHAPTER 2. SUNSET PRELUDE

1   W. H. Lehn, *Journal of the Optical Society of America 69, 776* (1979); also
    W. H. Lehn and I. I. Schroeder, *Polarforschung 49, 173* (1979).

CHAPTER 3. THE GREEN FLASH

1   D. J. K. O'Connell and C. Treusch, "The green flash," *Richershe Astron-
    omiche 4* (Specola Vaticana, Vatican City, 1958); also *The green flash and
    other low sun phenomena* (North Holland, 1958).
2   J. Murray, *Captain Back's narrative of the voyage of the H.M.S. Terror*
    (Murray, London, 1838), p. 191.
3   O'Connell and Treusch, "The green flash," p. 12.
4   The principal Vatican telescope used was a 40-cm (15.7-in.) refractor of
    160 cm (63 in.) focal length.

CHAPTER 4. THE EARTH'S SHADOW AND SUNSET PHENOMENA

1   C. F. Lummis, *A tramp across the continent* (Scribner, New York, 1892).
2   C. V. Rozenburg, *Twilight* (Plenum Press, New York, 1966).

CHAPTER 5. VOLCANIC ERUPTIONS

1   "Log of the William Besse," *Science 3, 702* (1884).
2   "Log of the Charles Bal," *Nature,* Jan. 10, 1884, p. 240; "Log of the
    Berbice," ibid., p. 242; "Log of the William Besse."
3   G. J. Symons (ed.), *The eruption of Krakatoa and subsequent phenomena*
    (Royal Society, London, 1888).
4   T. Simkin et al., *Volcanoes of the world* (Smithsonian Institution, Washing-
    ton, D.C., 1981).
5   Edward Wilson, *Diary of the discovery expedition to the Antarctic regions,
    1904,* A. Savours (ed.), (Blandford Press, Poole, Dorset, U.K., 1966), p.
    49. Distributed in the United States by Heritage Press.

Notes

CHAPTER 6. VOLCANIC TWILIGHTS

1   S. E. Bishop, *Science 3,* 216 (1884).
2   A. B. Meinel, M. P. Meinel, and G. E. Shaw, *Science 193,* 520 (1976).

CHAPTER 7. TWILIGHT SCIENCE

1   G. J. Symons (ed.), *The eruption of Krakatoa and subsequent phenomena* (Royal Society, London, 1888), p. 371.
2   A. B. Meinel and M. P. Meinel, *Science 155,* 189 (1967).

CHAPTER 8. BISHOP'S RING AND BLUE SUNS

1   G. J. Symons (ed.), *The eruption of Krakatoa and subsequent phenomena* (Royal Society, London, 1888), p. 232.
2   Ibid., p. 233.
3   R. Greenler, *Rainbows, Halos, and Glories* (Cambridge University Press, New York, 1980).
4   Symons, *Eruption of Krakatoa,* p. 200.
5   Ibid., p. 213.

CHAPTER 11. LIGHT OF THE NIGHT SKY AND THE AURORA

1   The aurora australis, the southern hemisphere equivalent of the aurora borealis, is in the direction of the south magnetic pole. It is seldom seen because few people live within the zone where auroras usually occur.

CHAPTER 12. CITY LIGHTS

1   A. B. Meinel, *Kitt Peak National Observatory Contributions 45,* 97 (1958).

CHAPTER 13. TWILIGHT ON THE PLANETS

1   J. J. Rinnelson and D. R. Criswell, *The Moon 10,* 121 (1974).

CHAPTER 14. CELESTIAL VISITORS

1   J. O. Langford with F. Gipson, *Big Bend, a homesteader's story* (University of Texas Press, Austin, 1952), pp. 119–21.

Name index

Abt, Helmut, 117
Ackerman, M., 74
Aitken, Robert, 20

Back, Captain, 20
Barents, Willem, 16
Bates, David, 105
Berkner, Lloyd, 105, 108
Bishop, S. E., 51, 79, 80, 158
Bok, Bart, 39
Brahe, Tycho, 146

Canright, Dick, 129
Castleman, A. N., Jr., 77
Cernan, Gene, 124
Chamberlain, Joseph, 108
Chandrasekhar, S., 107
Chang, Y. C., 85
Cortner, David, 112
Criswell, D. R., 124, 158

Edison, Thomas, 40
Edson, James B., 129, 130
Eric the Red, 17

Fan, Chung-Yan, 108
Franken, Peter, 84
Franklin, Benjamin, 150
Fuller, W. H., Jr., 75

Gartlein, Carl, 106
German, B., 17
Gipson, F., 158
Greenler, R., 158

Hale, George Ellery, 115, 117
Herzberg, Gerhard, 104, 105
Heustis, David, 112, 134
Hoag, Arthur, 96
Hoerlin, Herman, 114

Hu, Ningsheng, 13
Huxley, Julian, 47

Junge, C. E., 65

Kelvin, Lord, 20, 25
Kepler, Johannes, 16, 147
Keuper, P. F., 20
Kinnaird, Dick, 23
Knauth, L. P., 151

Lehn, W. H., 17, 157
Land, Edwin, 26, 83
Langford, J. O., 140, 158
Lechavallier, M., 74
Lippins, C., 74
Locanthi, Dorothy Davis, 145, 146
Lowe, D. R., 151
Lummis, Charles, 32, 157

Manowitz, B., 77
Mayall, N., 96
McCormick, M. P., 75
Meinel, A. B. and M. P., 158
Meldrum, C., 79
Mie, G., 84
Millikan, Robert, 116
Milon, Dennis, 139
Mossop, S. C., 65
Mukelwitz, H. R., 77
Mulliken, R. S., 108
Murray, J., 191

Nicolet, Marcel, 105

O'Connell, D. J. K., 19, 20, 21, 25, 26, 157

Pearson, Donald, 138
Pease, F. G., 116
Pettit, Edison, 105, 145

Name index

Richthofen, Baron von, 84
Rinnelson, J. J., 124, 158
Ritchey, G. W., 116
Roach, F. E., 103
Rozenberg, G. V., 36, 157

Schmidt, Harrison, 124
Schroeder, I. I., 17, 157
Schuster, Sir Arthur, 21
Shackleton, Sir Ernest, 16, 49
Simkin, T., 157
Smith, Harlan, 96
Swan, W., 20
Swings, Pol, 103
Symons, G. J., 157, 158

Treusch, C., 25, 26, 157
Tuve, Merle, 107

Van Allen, J., 107
Van Biesbrock, George, 115
Verne, Jules, 20
Volz, F. E., 77

Wilson, Edward, 49, 157
Winget, Jim, 129, 130
Wood, R. W., 103
Wright, Ernie, 129, 130

Zwicky, Fritz, 96, 147

Subject index

absorption, 9, 10, 15, 24, 59, 64, 77, 124
aerosol, 9, 29, 64–6, 74, 76, 77, 81, 121, 127
airglow, 122, 123
altitude of glow, 53
antisolar point, 34
ash, 57–9, 61, 65, 66, 74–6, 80, 82, 150
asteroid, 93, 142
astronaut, 123–6
atomic bomb, 114
aureole, 59, 81
aurora, 3, 106–14, 123, 147

Bali, 2
balloon observations, 73
Bishop's ring, 79–85
blue flash, 20, 23, 25
blue sun, *see* sun, blue

carbon dioxide, 150
Chinese lantern effect, 3, 14–16, 27
Christmas star, 147
city lights, 115–20
climate, 150–5
clouds
  ash, 82
  ice, 87
  man-made, 87
  noctilucent, 3, 7, 87–9
  smoke, 82
  water, 82
comet, 83, 133–6
  Halley, 135, 136
  Ikeya-Seki, 134, 135
  Kohoutek, 133
  West, 134, 135
contrail, 136
corona, 79, 95, 125
Crab nebula, 146

diffraction, 23, 81, 82
Doppler shift, 107, 111

dry fog, 42, 48, 150
dust, 9, 33, 56, 57, 64, 65, 81, 84, 87, 93, 95, 124, 127, 135

ecliptic, 5, 91, 93, 124, 125
electromagnetic pulse (EMP), 113
equinox, 6, 93
eye, 11, 12, 21, 23–5, 36, 73, 83, 92

gegenschein, 4, 95–8
geometry of sunset, 6
glow, 40, 51–61, 63–73, 97, 118, 120
gnommon, 68
green flash, 3, 12, 19–27, 83
green sun, *see* sun, green
greenhouse effect, 2, 151

halo, 79, 81
haze, 84
heiligeschein, 95
hydrogen, 106, 111, 113

intensity, 11
interplanetary material, 95
Io, 130

Junge layer, 65, 66
Jupiter, 95, 130, 135, 143

latitude, 5
lidar, 56, 74
lights, city, 115–20
little ice age, 48

magnetic
  field, 114
  force, 111, 114
  lines, 106, 114
  pole, 108, 111
Mars, 95, 127, 143
Mercury, 95, 143

Subject index

meteor, 87, 89, 95, 124, 125, 137–42
  crater, 143
Mie effect, 84
Milky Way, 98
mirage, 16
moon, 14, 31, 81, 120, 123–6
  blue, 52
  green, 52, 83
moonrise, 122

night
  astronomical, 36, 38
  civil, 36
  light of the night sky, 37, 101–5, 117
nitrogen, 108, 111, 123
noctilucent clouds, 3, 7, 87–9
nova, 144–7
  Puppis, 145
Novaya Zemlya mirage, 16, 17
nuclear detonation, 113

oblateness, 14, 121–3
observatories,
  Harvard, 134, 145
  Kitt Peak, 31, 52, 103, 108, 112, 117, 134, 138, 139
  Lick, 117
  Lowell, 129
  McDonald, 115
  Mount Palomar, 117
  Mount Wilson, 116, 117, 145, 147
  Purple Mountain, 85
  Table Mountain, 112, 129
  Whipple, 112
  Yerkes, 52, 101, 104, 106, 111, 116, 117
OH, 105, 123
Orion, 102, 103
oxygen, 104, 105, 108, 114, 123, 127, 151
ozone, 9, 10, 65, 66, 105

photometric observations, 70, 71
polarize, 95
protons, 107
purple light, 36, 64, 111

rainbow, 81
Rayleigh scattering, 10, 121
rays
  auroral, 111, 114
  blue, 53
  Buddha, 33
  cloud, see shadow
reddening, reddened, 9, 15, 52, 55, 84, 127
refraction, 12–17, 21, 23, 25, 26, 82
rocket, 88
Royal Society of London, 47

satellite, 74, 136
  Apollo, 123, 125, 126
  Echo, 1, 2, 137
  Skylab, 121, 122, 123, 137
  Soyuz, 137
  Surveyer, 123
  Venera, 130
  Voyager 1, 131
  Voyager 2, 128, 131
Saturn, 135, 143
scattering, 10
scintillation, 98
"seeing," astronomical, 99
shadow, 3, 12, 29–34, 52, 71, 72, 98, 121, 127
Sirius, 98
smoke, 82, 84
solar collector, 12
solstice, 5, 8
speckle interferometry, 100
spectrum, 64
star
  Kepler's, 147
  Tycho's, 147
sulfur dioxide, 65, 66, 150
sulfuric acid, 65, 66, 76
sun
  aureole, 58
  blue, 47, 48, 79, 83–5
  green, 47, 48, 79, 83–5
  flattened, 121, 122
  limb darkening, 122
sunlight extinction, 76
sunrise, 5, 8, 9, 16, 21, 35, 121, 125
sunset, 1–3, 8, 9, 17, 21, 34–6, 39, 40, 51–61, 121, 125, 149
supernova, 144–7

tangent, 30, 67, 69, 70, 122
Takli Makan desert, 79
Titan, 128
twilight, 1, 2, 6, 9, 36–8, 47, 48, 51–61, 63–78
  lunar, 124
  on planets, 121–31
twinkling, 98

unidentified flying object (UFO), 98, 137

Van Allen belts, 3, 114
vapor trail, 136
Venus, 22, 128, 150
vignetting, 92, 96, 102
visibility, 11
volcano, 39–50
  Agung, 2, 31, 49, 51–61, 63, 65, 66, 71, 74, 76, 77

volcano (*cont.*)
Anak Krakatau, 47
Asama, 48
Awu, 49
Belusan, 77
Bezymia, 41
Cotopaxi, 49
El Chichón (Chichonal), 31, 42, 49, 58–61, 66, 78, 81
El Fuego, 48, 49, 55, 59, 66, 75, 76, 81, 83
Etna, 41
Fernandia, 48
Fuji, 41
Galunggung, 61
Guagua Pichincha, 48
Hekla, 48, 57
Irazu, 49, 77
Karthala, 77
Katmai, 41, 76–8
Kelut, 77
Krakatoa, 2, 19, 31, 32, 39–40, 43–7, 49, 51, 53, 63, 76, 80, 83, 84, 143
Mayon, 48, 77
Mazama, 42
Merapi, 77
Peleé, 49
Saint Augustine, 41, 56
Saint Helens, 41, 42, 50, 57, 74
Saint Vincent, 48
Santa Maria, 77
Santorin, 43
Skaptar Jokull, 48
Soufrière, 49
Surtsey, 76
Taal, 77
Tambora, 19, 48, 149
Vesuvius, 41

water vapor, 10

zodiacal light, 3, 91–7, 101, 124, 125